T0183857

Communications
in Computer and Information Science   1054

Commenced Publication in 2007
Founding and Former Series Editors:
Phoebe Chen, Alfredo Cuzzocrea, Xiaoyong Du, Orhun Kara, Ting Liu,
Krishna M. Sivalingam, Dominik Ślęzak, Takashi Washio, and Xiaokang Yang

More information about this series at http://www.springer.com/series/7899

Muhammad Younas · Irfan Awan ·
Salima Benbernou (Eds.)

Big Data Innovations
and Applications

5th International Conference, Innovate-Data 2019
Istanbul, Turkey, August 26–28, 2019
Proceedings

 Springer

Editors
Muhammad Younas
School of ECM
Oxford Brookes University
Oxford, UK

Irfan Awan
Department of Informatics
University of Bradford
Bradford, UK

Salima Benbernou
Universite Paris Descartes
Paris, France

ISSN 1865-0929 ISSN 1865-0937 (electronic)
Communications in Computer and Information Science
ISBN 978-3-030-27354-5 ISBN 978-3-030-27355-2 (eBook)
https://doi.org/10.1007/978-3-030-27355-2

This Springer imprint is published by the registered company Springer Nature Switzerland AG
The registered company address is: Gewerbestrasse 11, 6330 Cham, Switzerland

Preface

Welcome to the proceedings of the 5th International Conference on Big Data Innovations and Applications (Innovate-Data 2019), which was held during August 26–28, 2019, in Istanbul, Turkey. The Innovate-Data 2019 conference was co-located with the International Conference on Future Internet of Things and Cloud (FiCloud-2019), the International Conference on Mobile Web and Intelligent Information Systems (MobiWis-2019), the International Conference on Deep Learning and Machine Learning in Emerging Applications (DEEP-ML 2019), and different workshops and symposia.

Big data has become a new trend in modern information technology and has been transforming various domains and disciplines including business, finance, science, education, the public sector, health care, life style and society as a whole. For instance, businesses can exploit big data in order to improve customer relationships and to serve customers in an efficient and effective way. Big data tools can also help in cost and time savings as they can analyze data efficiently and make quick decisions.

The concept of big data revolves not only around the size but how the data can be efficiently collected, processed, and analyzed in order to achieve the desired goals. The aim of the Innovate-Data conference is to solicit research and development work in order to further improve the underlying models, methods, and tools of big data and to find new ways of using big data in current and future applications. The Innovate-Data 2019 conference included interesting and timely topics such as: big data storage, representation, and processing; data engineering and design; big data security, privacy, and trust; big data models, infrastructure, and platforms; visualization of big data; big data analytics and metrics; and the overall applications and innovations of big data and related technologies.

In response to the call for papers, Innovate-Data 2019 received submissions from different countries across the world. Each paper was reviewed by multiple reviewers. Based on the reviews, 16 papers were accepted for the technical program, that is, an acceptance rate of 33%. The accepted papers address interesting research and practical issues related to big data processing and applications, big data analytics, big data security, privacy and trust, machine/deep learning and big data, and the use of big data in innovative applications.

We express our thanks to the members of the Organizing and Program Committees for helping in the organization and the review process of the research papers. Their timely and useful feedback to authors of the submitted papers is highly appreciated. We would also like to thank all authors for their contributions to Innovate-Data 2019. Putting together such an interesting technical program would not have been possible without the support from the technical committees and the authors.

We sincerely thank Prof. William Knottenbelt (General Chair) for his help and support. We would like to thank Dr. Filipe Portela (Workshop Coordinator), Dr. Fang-Fang Chua (Publicity Chair), Dr. Lin Guan (Journal Special Issue Coordinator), and Prof. Irfan Awan (Publication Chair). We sincerely appreciate the contributions of the local organizing chairs, Dr. Perin Ünal, Teknopar, Dr. Tacha Serif, and Dr. Sezer Gören Uğurdağ, for their help and support in the organization of the conference.

Innovate-Data 2019 and the co-located conferences had joint keynote and invited sessions in which interesting talks were delivered. We sincerely thank the speakers Prof. Pierangela Samarati (Università degli Studi di Milano, Italy), Mr. Gökhan Büyükdığan (Arçelik A.S., Turkey), and Dr. Soumya Kanti Datta (EURECOM, France).

Our sincere thanks also go to the Springer CCIS team for their valuable support in the production of the conference proceedings.

August 2019   Salima Benbernou
                                                             Muhammad Younas

Organization

Innovate-Data 2019 Organizing Committee

General Chair

William Knottenbelt — Imperial College London, UK

Program Co-chairs

Salima Benbernou — Université Paris Descartes, France
Muhammad Younas — Oxford Brookes University, UK

Local Organizing Co-chairs

Perin Ünal — Teknopar, Turkey
Sezer Gören Uğurdağ — Yeditepe University, Turkey
Tacha Serif — Yeditepe University, Turkey

Publication Chair

Irfan Awan — University of Bradford, UK

Workshop Coordinator

Filipe Portela — University of Minho, Portugal

Publicity Chair

Fang-Fang Chua — Multimedia University, Malaysia

Journal Special Issue Coordinator

Lin Guan — Loughborough University, UK

Program Committee

Abdelmounaam Rezgui — Illinois State University, USA
Abdulsalam Yassine — Lakehead University, Canada
Afonso Ferreira — CNRS - Institut de Recherches en Informatique de Toulouse, France
Ahmed Awad — Cairo University, Egypt

Ahmed Zouinkhi	National Engineering School of Gabes, Tunisia
Akiyo Nadamoto	Konan University, Japan
Allaoua Chaoui	University Mentouri Constantine, Algeria
Amin Beheshti	Macquarie University, Australia
Armin Lawi	Hasanuddin University, Indonesia
Amr Magdy	University of California, Riverside, USA
Aris Gkoulalas-Divanis	IBM Watson Health, Cambridge, MA, USA
Athena Vakali	Aristotle University of Thessaloniki, Greece
Ayman Alahmar	Lakehead University, Canada
Cándido Caballero-Gil	Universidad de La Laguna, Spain
Dimitri Plexousakis	University of Crete, Greece
Dimka Karastoyanova	University of Groningen, The Netherlands
Domenico Talia	University of Calabria, Italy
Emad Mohammed	University of Calgary, Canada
Ernesto Damiani	University of Milan, Italy
Fanny Klett	German Workforce ADL Partnership Laboratory, Germany
Faouzi Alaya Cheikh	Norwegian University of Science and Technology, Norway
Federica Paci	University of Trento, Italy
Giuseppe Di Modica	University of Catania, Italy
George Pallis	University of Cyprus, Cyprus
Guy De Tré	Ghent University, Belgium
Hesham Hallal	Fahad Fahad Bin Sultan University, Saudi Arabia
Hiroaki Higaki	Tokyo Denki University, Japan
Ibrahim Korpeoglu	Bilkent University, Turkey
Irena Holubova	Charles University in Prague, Czech Republic
Iván Santos-González	Universidad de La Laguna, Spain
Jezabel M. Molina-Gil	Universidad de La Laguna, Spain
Jorge Bernardino	Polytechnic Institute of Coimbra - ISEC, Portugal
Kazuaki Tanaka	Kyushu Institute of Technology, Japan
Khouloud Boukadi	University of Sfax, Tunisia
Mohamed Boukhebouze	CETIC Research Center, Belgium
Morad Benyoucef	University of Ottawa, Canada
Mourad Khayati	University of Fribourg, Switzerland
Mourad Ouziri	Université Paris Descartes - LIPADE, France
Muhammad Rizwan Asghar	The University of Auckland, New Zealand
Orazio Tomarchio	University of Catania, Italy
Pino Caballero-Gil	DEIOC, University of La Laguna, Spain
Rachid Benlamri	Lakehead University, Canada
Radwa El Shawi	University of Tartu, Estonia
Rafael Santos	Brazilian National Institute for Space Research, Brazil
Raghava Mutharaju	IIIT-Delhi, India
Raj Sunderraman	Georgia State University, USA
Ridha Hamila	Qatar University, Qatar
Saad Harous	UAE University, UAE

Sadiki Tayeb   International University Rabat, Morocco
Sebastian Link   The University of Auckland, New Zealand
Shantanu Sharma   University of California, Irvine, USA
Stephane Bressan   National University of Singapore, Singapore
Toshihiro Yamauchi   Okayama University, Japan
Tengku Adil   University Technology of MARA, Malaysia
Vasilios Andrikopoulos   University of Groningen, The Netherlands
Yacine Atif   Skövde University, Sweden
Zaki Malik   Eastern Michigan University, USA
Zuhair Khayyat   KAUST, Saudi Arabia

Contents

Advances in Big Data Systems

Enabling Joins over Cassandra NoSQL Databases 3
  Haridimos Kondylakis, Antonis Fountouris, Apostolos Planas,
  Georgia Troullinou, and Dimitris Plexousakis

MapReduce Join Across Geo-Distributed Data Centers 18
  Giuseppe Di Modica and Orazio Tomarchio

Mobile Device Identification via User Behavior Analysis 32
  Kadriye Dogan and Ozlem Durmaz Incel

A SAT-Based Formal Approach for Verifying Business
Process Configuration 47
  Abderrahim Ait Wakrime, Souha Boubaker, Slim Kallel,
  and Walid Gaaloul

Machine Learning and Data Analytics

TSWNN+: Check-in Prediction Based on Deep Learning
and Factorization Machine 65
  Chang Su, Ningning Liu, Xianzhong Xie, and Shaowen Peng

Deep Learning Based Sentiment Analysis on Product Reviews on Twitter . . . 80
  Aytuğ Onan

A Cluster-Based Machine Learning Model for Large Healthcare
Data Analysis 92
  Fatemeh Sharifi, Emad Mohammed, Trafford Crump,
  and Behrouz H. Far

Satire Detection in Turkish News Articles: A Machine
Learning Approach 107
  Mansur Alp Toçoğlu and Aytuğ Onan

Big Data Innovation and Applications

Committee of the SGTM Neural-Like Structures with Extended Inputs
for Predictive Analytics in Insurance 121
  Roman Tkachenko, Ivan Izonin, Michal Greguš ml., Pavlo Tkachenko,
  and Ivanna Dronyuk

Game Analytics on Free to Play   133
  Robert Flunger, Andreas Mladenow, and Christine Strauss

A Decentralized File Sharing Framework for Sensitive Data.   142
  Onur Demir and Berkay Kocak

The Information System for the Research in Carotid Atherosclerosis   150
  Jiri Blahuta and Tomas Soukup

Security and Risk Analysis

Big Data Analytics for Financial Crime Typologies.   165
  Kirill Plaksiy, Andrey Nikiforov, and Natalia Miloslavskaya

Development of a Model for Identifying High-Risk Operations
for AML/CFT Purposes.   179
  Pavel Y. Leonov, Viktor P. Suyts, Oksana S. Kotelyanets,
  and Nikolai V. Ivanov

Monitoring System for the Housing and Utility Services Based
on the Digital Technologies IIoT, Big Data, Data Mining, Edge
and Cloud Computing   193
  Vasiliy S. Kireev, Pyotr V. Bochkaryov, Anna I. Guseva,
  Igor A. Kuznetsov, and Stanislav A. Filippov

K-Means Method as a Tool of Big Data Analysis in Risk-Oriented Audit . . .   206
  Pavel Y. Leonov, Viktor P. Suyts, Oksana S. Kotelyanets,
  and Nikolai V. Ivanov

Author Index   217

Advances in Big Data Systems

Enabling Joins over Cassandra NoSQL Databases

Haridimos Kondylakis[1,2], Antonis Fountouris[1,2], Apostolos Planas[1,2],
Georgia Troullinou[1,2(✉)], and Dimitris Plexousakis[1,2]

[1] Institute of Computer Science, FORTH, Heraklion, Crete, Greece
{kondylak, troulin}@ics.forth.gr
[2] Computer Science Department, University of Crete, Heraklion, Crete, Greece

Abstract. Over the last few years, we witness an explosion on the development
of data management solutions for big data applications. To this direction,
NoSQL databases provide new opportunities by enabling elastic scaling, fault
tolerance, high availability and schema flexibility. Despite these benefits, their
limitations in the flexibility of query mechanisms impose a real barrier for any
application that has not predetermined access use-cases. One of the main reasons
for this bottleneck is that NoSQL databases do not directly support joins. In this
paper, we propose a data management solution, designed initially for eHealth
environments, that relies on NoSQL Cassandra databases and efficiently supports joins, requiring no set-up time. More specifically, we present a query
optimization and execution module, that can be placed, at runtime, on top of any
Cassandra cluster, efficiently combining information from different column-
families. Our optimizer rewrites input queries to queries for individual column-
families and considers two join algorithms implemented for the efficient execution of the requested joins. Our evaluation demonstrates the feasibility of our
solution and the advantages gained, compared to the only solution currently
available by DataStax. To the best of our knowledge, our approach is the first
and the only available open source solution allowing joins over NoSQL Cassandra databases.

1 Introduction

The development of new scientific techniques and the emergence of new high
throughput tools have led to a new information revolution. The nature and the amount
of information now available open directions of research that were once in the realm of
science fiction. During this information revolution the data gathering capabilities have
greatly surpassed the data analysis techniques, making the task to fully analyze the data
at the speed at which it is collected a challenge.

Traditionally, Relational Database Management Systems (RDBMS) were
employed for data management. However, the aforementioned growth in the number of
users, applications and volume of data, dramatically changed the trends and the
computing landscape and led to the rise of NoSQL databases. While non-relational
databases date back to the late 1960s, after the developments by Google (Big Table [1])
and Amazon (DynamoDB [2]) the trend again gained traction. In nowadays, even

© Springer Nature Switzerland AG 2019
M. Younas et al. (Eds.): Innovate-Data 2019, CCIS 1054, pp. 3–17, 2019.
https://doi.org/10.1007/978-3-030-27355-2_1

traditional RDBMS vendors such as Oracle and IBM are focusing on NoSQL products and many companies like Facebook, Twitter, Amazon, Reddit, Netflix etc. are using them for big data applications. Among the reasons for the rapid adoption of NoSQL databases is that they scale across a large number of servers by horizontal partitioning data items, they are fault tolerant and achieve high write throughput, low read latencies and schema flexibility. To achieve all these benefits, the main idea is that you have to denormalize your data model and avoid costly operations in order to speed up the database engine. NoSQL databases were initially designed to support only single-table queries and explicitly excluded the support for join operations allowing applications [3] to implement such tasks.

However, the latest years more and more applications require the efficient combination of information from multiple sources of information [4, 5]. To this direction, many approaches have already appeared, implementing operators similar to join, based on Map-Reduce, such as rank-join queries [6] and set-similarity joins [7]. Rank-join queries usually try to find the most relevant documents for two or more keywords, by executing a rank-join over corresponding row lists. On the other hand, set-similarity joins are those that try to find similar pairs of records instead of exact ones. However, both these approaches execute joins at the application level using Map-Reduce implementations and the joins implemented do not focus on an exact matching of the joined tuples. Other approaches, try to offer SQL engines over key-value stores (Partiqle [8], Google App Engine [9], Google Megastore [10] etc.) introducing a data model between the query engine and the key-value stores. However, we are interested in an approach that (a) it would be able to *directly use the existing sources* (b) without *modifying the existing infrastructure* and (c) *without introducing any additional modelling overhead (models, mappings etc.) to the end-user.*

This emerging need has also been recognized by DataStax, the biggest vendor of Cassandra NoSQL commercial products, which introduced a join-capable ODBC driver. The company claims that Cassandra can now perform joins just as well as relational database management systems [11]. However, information is missing for both the specific join implementation algorithms and the optimization techniques used and evaluation results have not been presented extensively. We have to note that the driver is commercial, was first made available on March 2015, the source code is not available, and a fee is required for using it.

To fill these gaps, in this paper we present a novel model, capable of executing joins over Cassandra NoSQL databases with *completely no setup overhead for the end-user*. More specifically our contributions are the following:

1. We present an effective and efficient plug-and-play solution named *CassandraJoins* for querying and combining NoSQL data with no setup overhead.
2. Our solution is implemented as a "Query Optimization & Execution" module that can be placed on top of any Cassandra cluster, enabling effective execution of join operations.
3. This module includes a simple, yet efficient optimizer for executing joins over Cassandra NoSQL databases, implementing and extending two join algorithms: (a) the Index-Nested Loops and (b) the Sort-Merge join algorithms.

4. We describe the corresponding algorithms and their extensions and we show their computational complexity.
5. Finally, we show experimentally that our approach strictly dominates the available commercial implementation in terms of execution time, in most cases by one order of magnitude.

CassandraJoins was first implemented within the iManageCancer H2020 project [12] in order to facilitate ontology-based access over NoSQL databases through mappings [13, 14], and then evolved through the BOUNCE H2020 project. The first seeds of the approach were initially presented as a conference poster [15]. Out of this single poster more than five commercial companies have already showed interest for acquiring our driver, identifying the potential impact of our solution. As soon as the main paper of the approach is published, a java package will be released as an open source package providing an API for performing joins over Cassandra NoSQL databases. To the best of our knowledge, *our approach is the first non-commercial implementation allowing joins over NoSQL* Cassandra databases, dominating the existing commercial one.

The remaining of this paper is structured as follows: In Sect. 2, we present some background on NoSQL databases for better understanding our contributions. Section 3 presents the optimizer and the two implemented join algorithms and Sect. 4 demonstrates our implementation and presents our evaluation results. Finally, Sect. 5 concludes this paper and presents an outlook for further work.

2 Background

In this section, we briefly describe the basic ideas behind NoSQL databases [18] and we focus on the special characteristics of Cassandra that should be considered when implementing joins in such a diverse environment. NoSQL databases were designed to provide solutions for large volumes of data. Motivations for this approach include, horizontal scaling, simplicity of design and finer control over availability. Moreover, NoSQL databases are designed to handle all type of failures, which are no longer considered as exceptional events but as eventual occurrences. NoSQL databases use looser consistency models than traditional relational databases. When the amount of data is large the ACID principle (atomic – consistent – isolated – durable) that traditional databases adopt is hard to attain [19]. That's why NoSQL focuses on the BASE principle:

- **BA**sically Available: All data are distributed and even if there is a failure the system continues to work.
- **S**oft state: There is no consistency guarantee.
- **E**ventually consistent: System guarantees that even when data are not consistent, eventually they will be.

It is important to note that BASE still follows the CAP (consistency – availability – partition tolerance) theorem and if the system is distributed, two of three guarantees must be chosen depending on the database purpose. BASE is more flexible than ACID and the big difference is about consistency. If consistency is crucial, relational databases may be better solution but when there are hundreds of nodes in a cluster, consistency becomes very hard to accomplish.

There have been various approaches to classify NoSQL databases and the most common classification is according to the data models they use [20]. In key-value stores all data are stored as a set of key and value, in document stores these key-values are transformed into documents whereas in graph databases data are represented as graphs. In another type of NoSQL databases, called column-family stores, data are structured in columns that may be countless. This is the type that resembles most closely to relational databases. In this paper, we are going to focus on these types of NoSQL databases and more specific on Cassandra NoSQL databases [21].

Cassandra is a NoSQL database developed by Apache Software Foundation. It uses a hybrid model between key-value and column-oriented database. Data is structured in columns and organization of data consists of the following blocks

- **Column:** Represents units of data identified by key and value
- **Super-column:** They are columns grouped by information columns
- **Column-family:** A set of structured data similar to relation database table, constituted by a variety of super columns.

The structure of the database is defined by super-columns and column-families. In this paper, the term column-family and table will be used interchangeably although it is not exactly the same as previously explained. New columns can be added whenever necessary and data access can be done by indicating column-family, key and column in order to obtain values.

Cassandra can handle billions of columns and millions of operations per day whereas data can be distributed efficiently all over the world. When a node is added or removed, all data are automatically distributed over other nodes and the failed node can be replaced with no downtime. Two important features of Cassandra is durability and indexing. In Cassandra, the primary index for a column family is the index of its row keys. Replication on the other hand can be synchronous and asynchronous and each node maintains all indexes of the tables it manages. Rows are assigned to nodes by the cluster-configured *partitioner* and the keyspace-configured *replica placement strategy*.

Primary keys in Cassandra are considered as *partition keys*. When we have composite primary keys the first is considered to be the *partition key* and the rest the *clustering keys*. The partition is the unit of replication in Cassandra. Partitions are distributed by hashing and a ring is used to locate the nodes that store the distributed data. Cassandra would generally store information on different nodes, and to find all information requested usually Cassandra has to visit different nodes. To avoid these problems, each node indexes its own data. Since each node knows what ranges of keys each node manages, requested rows can be efficiently located by scanning the row indexes only on the relevant replicas. With randomly partitioned row keys (the default in Cassandra), row keys are partitioned by their MD5 hash and cannot be scanned in order like traditional b-tree indexes. However, it is possible to page through non-ordered partitioned results. Cassandra uses the indexes to pull out the records in question. An attempt to filter the data before creating an index would fail because the operation would be very inefficient[1]. In addition, although initially Cassandra didn't support secondary indexes as of January 2011 these indexes were allowed. Cassandra implements secondary indexes as a hidden table, separate from the table that contains the values being indexed.

All stored data can be easily manipulated using Cassandra Query Language (CQL) which is based on the widely used SQL. CQL can be thought as an SQL fragment with the following restrictions over the classical SQL:

- R1: Joins are now allowed
- R2: You cannot project the value of a column without selecting first the key of the column.
    - Every select query requires that you restrict all partition keys
    - Select queries restricting a clustering key have to restrict all the previous clustering keys in order
    - Queries that don't restrict all partition keys and any possibly required clustering keys, can run only if they can query secondary indexes
    - To run a query including more than two secondary indexes, Cassandra requires that *allow filtering* is used in the query to show that you really want to do it. All Cassandra queries that require allow filtering run extremely slow and Cassandra's recommendation is to avoid running them.
    - Tables can be stored sorted by clustering keys. This is the only case in which you are allowed to run range queries and order by clauses
- R3: Unlike the projection in a SQL SELECT, there is no guarantee that the results will contain all of the columns specified because Cassandra is schema-optional. An error does not occur if you request non-existent columns.
- R4: Nested queries are not allowed, there is no "OR" operator and queries that select all rows of a table are extremely slow

CQL statements change data, look up data, store data or change the way data is stored. A select CQL expression selects one or more records from Cassandra column family and returns a result-set of rows. Similarly to the SQL each row consists of a row key and a collection of columns corresponding to the query.

3   Query Optimization and Execution

Our query optimization and execution module can be placed on top of any Cassandra cluster and is composed of the following components:

Rewriter. The rewriter accepts the CQL query containing joins and creates the queries for accessing each individual column-family/tables. For example, assume that the column-families *movies* and *producedBy* are available and Q is issued by the user:

```
CREATE TABLE movie (movieId varchar PRIMARY KEY, movieTitle
varchar, language varchar, year int)
```

---

[1] http://docs.datastax.com/en/cassandra/2.1/cassandra/dml/dml_index_internals.html.

```
CREATE TABLE producedBy (id in PRIMARY KEY, movieId int,
companyName text)
```

Q: SELECT movieTitle, companyName FROM movie, producedBy
WHERE movie.movieId=producedBy.movieId AND year=2015;

Q is parsed by the rewriter and the following two queries *Q1* and *Q2* are generated:

Q1:   SELECT movieId, movieTitle FROM movie WHERE year=2015
Q2:   SELECT movieId, companyName FROM producedBy

Table 1. Heuristic rules

H1	If input query don't restrict all partition keys, generate appropriate secondary indexes
H2	If two column-families should be joined, and only one of them has a partition index on the joined field, then use the Index-Nested Loops join algorithm, using the indexed relation as the inner relation in the corresponding for loop
H3	If two column-families should be joined, and they both have a partition index on the joined fields, then use the sort-merge join algorithm
H4	If two column-families should be joined, and they don't have a partition index on the joined fields, sort them and then use the sort-merge join algorithm

Planner. This component tries to plan the execution of the individual queries, as constructed by the rewriter, using the following heuristics, shown also in Table 1.

First it identifies the available indexes on the queried column-families and tries to comply with R2. For example, if the queries don't restrict all partition keys they can only run if there are available secondary indexes on these keys. To satisfy this restriction the planner automatically generates secondary indexes on the required fields. In our running example, a secondary index will automatically be generated by the planner component to the *producedBy.movieID* column.

Besides trying to comply with all Cassandra restrictions the planner identifies which one of the two join implementations should be used for executing the join, i.e. the Index-Nested Loops and the Sort-Merge join algorithms. For example, when two column-families should be joined, if only one of them has an index on the joined field, we could read all rows from the non-indexed one and then use the index for searching the indexed column-family. On the other hand, when both column-families are sorted on the join column (i.e. when the column is the partition key) the Sort-Merge join algorithm would be faster and is used.

Combiner. This component sends the individual queries to the Cassandra cluster in order to be answered, combines the results according to the generated execution plan and returns to the user the results. We have to note that our module focuses on generating the queries that will access the individual column families and optimizing the combination of the returned results. The individual queries generated are subsequently further optimized and executed by the Cassandra query engine.

3.1 Index-Nested Loops

Next, we describe in detail the two implemented join algorithms. The simplest join algorithm when an index is available is Index-Nested loops. Our implementation is shown in Fig. 1. The algorithm takes advantage of the index by making the indexed relation to be the inner relation. For each tuple $r \in R$, we use the index to retrieve the matching tuples of S.

Algorithm 1: ExtendedIndexNestedLoops (R, S, '$ri{=}{=}sj$' || 'ri CONTAINS sj')
Input: R, S the input column-families, ri, sj the joined fields
Output: The joined result
1. $result := \{\}$;
2. foreach tuple $r \in R$
3. foreach tuple $s \in S$ where $ri{=}{=}sj$ || ri CONTAINS sj
4. add $\langle r, s \rangle$ to the *result;*
5. Return *result;*

Fig. 1. The extended Index-Nested Loops algorithm.

The differences with the traditional Index-Nested Loops algorithm 23 are the following: (a) Our algorithm supports a new join paradigm using CONTAINS over collection sets (explained below); (b) In our case, to retrieve each tuple $s \in S$ a new query is constructed and sent to the Cassandra database.

In our running example for instance, based on $Q2$, the following set of queries will be used for retrieving the selected tuples from the *producedBy* column-family:

Qi: SELECT movieId, companyName FROM producedBy WHERE producedBy.movieId=<movieid_i> where movieId_i $\in$ movie. movieId as they are retrieved by Q1.

Next we analyze the cost of the corresponding algorithm using the disk access model [22], which measures the runtime of an algorithm in terms of disk pages transferred. In Table 2 the terms used for this analysis are presented. If we assume that R needs M pages to be stored, there are tR number of tuples selected from R and that the cost for performing an index scan is K then the total I/Os required for executing the algorithm is $M + K * tR$ I/Os [23]. Since Cassandra is using hash indexes, usually accessing a single tuple using the index requires on average 2 I/Os.

We have to note that the algorithm can only run on Cassandra versions supporting automatic paging (after 2.0) to allow retrieving all necessary tuples of the outer column family (line 2). In addition, after the version 2.1, Cassandra allows indexing collections of elements (maps, sets and lists) with a new type of index. Queries can benefit from these indexes by asking for collections containing specific values. Our Index-Nested Loops join algorithm can handle collection indexes as well, allowing queries that join a column from one column-family with a collection from another column-family. Consider for example the following two tables:

Table 2. Table of terms

M	Pages of R relation, pR tuples per page
tR	Number of tuples selected from relation R
psR	Pages occupying of the selected tuples from R relation
N	Pages of S relation, pS tuples per page
tS	Number of tuples selected from relation S
psS	Pages occupying the selected tuples from S relation
K	I/Os required for retrieving a single tuple using a primary index

```
CREATE TABLE elements (id int PRIMARY KEY, description text)
CREATE TABLE collections (id int PRIMARY KEY, s set<int>,
data varchar)
```

Consider also a collection index at the column s and the following query:

Q3: `SELECT * FROM elements, collections WHERE elements.id=5 AND collections.s CONTAINS elements.id`

Obviously, the column s of the column-family *collections* contains many values that should to be joined with the *elements.id* field. In our case, the *rewriter* will split *Q3* to the *Q4* and *Q5* shown below.

Q4: `SELECT * FROM elements WHERE elements.id=5`
Q5: `SELECT * FROM collections`

Then the Index-Nested Loops algorithm will use s as the inner relation since a collection index there. As already explained, the algorithm instead of using Q5, it will use Q6 for retrieving efficiently the joined rows from the *collections* column-family.

Q6: `SELECT * FROM collections WHERE s CONTAINS 5`

3.2 Sort-Merge Join

As already described, the Sort-Merge join algorithm is preferred by the optimizer when both relations that are provided as input are sorted. The corresponding implementation is shown in Fig. 2. As both column-families are sorted on the joining field, the algorithm scans one time each relation to produce the joined result. The cost of the algorithm is $M + N$ I/Os, where M is the number of pages occupied by R and N is the number of pages occupied by S [23].

Besides already sorted relations, the algorithm can be also used with unsorted ones. However, in this case sorting them is required before starting joining tuples (lines 3–4).

Sorting on a non-indexed field would require (a) either the creation of a secondary index and then accessing all tuples in the column-family or (b) selecting all pages, materializing them, sorting using external sort and then performing sort-merge join.

The former has a cost of $M + N$ I/Os for reading the pages of the relations, for indexing (neglecting the cost for writing the secondary index which usually is really small) adding $tR * K$ I/Os for accessing tuples selected from relation R and $tS*K$ I/Os for accessing tuples selected from relation S, plus $psR + psS$ I/Os for performing the join on the resulted tuples. In total $M + N + (tR * K) + (tS * K) + psR + psS$ I/Os.

Algorithm 2: ExtendedSortMergeJoin (R, S. '*ri==sj*')
Input: R, S the input column-families, *ri, sj* the joined fields
Output: The joined result
1. *result:=* {};
2. if *R* not sorted on attribute *i*, sort it; if *S* not sorted on attribute *j*, sort it;
3. *Tr* = the first tuple in *R; Ts* = the first tuple in *S; Gs* = the first tuple in *S;*
4. while *Tr* ≠ *eof* and *Gs* ≠ *eof* do {
5.   while *Tri* < *Gsj* do
6.   *Tr* = next tuple in *R* after *Tr;*
7.   while *Tri* > *Gsj* do
8.   *Gs* = next tuple in *S* after *Gs;*
9..   *Ts = Gs;*
10.   while *Tri* == *Gsj* do {
11.   *Ts = Gs;*
12.   while *Tsj* == *Tri* do {
13.   add ⟨ *Tr,Ts* ⟩ to *result;*
14.   *Ts* = next tuple in *S* after *Ts;*}
15.   *Tr* = next tuple in *R* after *Tr;* }
16.   *Gs = Ts;* }
17. Return *result;*

Fig. 2. The algorithm for index nested loops

On the other hand for the latter method, $M + N$ I/Os are required for the initial selection of all pages from the two relations, plus approximately four passes over the resulted data from the two relations for sorting them using external sort[2] $4(psR + psS)$ I/Os plus $psR + psS$ I/Os for joining them. In total $M + N + 5(Rsel_p + Ssel_p)$ I/Os. Usually, $4(Rsel_p + Ssel_p) < K * (Rsel_t + Ssel_t)$ and as such the latter method is preferred.

As an example consider the following two ordered tables and query Q7:

```
CREATE TABLE e1 (event_type text, price int, PRIMARY KEY
(event_type, price)) WITH CLUSTERING ORDER BY (price DESC);
CREATE TABLE e2 (event_type text, price int, PRIMARY KEY
(event_type, price)) WITH CLUSTERING ORDER BY (price DESC);
```

Q7: SELECT * FROM e1, e2 WHERE e1.price = e2.price;

---

[2] External sorting comprises two phases: partitioning and merging. The partitioning phase involves scanning the raw file in chunks that fit in main memory, sorting each chunk in main memory, and flushing it to secondary storage as a sorted partition. This amounts to two passes over the data. The merging phase involves merge-sorting all the different partitions into one contiguous sorted order, using one input buffer for each partition and one output buffer for the resulting sorted order. Thus, the merging phase amounts to two additional passes over the data, and so external sorting involves overall four passes over the data. In fact this condition only holds as long as M > N [23].

The query will be split to the following queries by the rewriter:

Q8: SELECT * FROM e1
Q9: SELECT * FROM e2

Then the planner will identify that the joined column-families are already ordered so the Sort-Merge Join algorithm will be preferred. The algorithm will be executed then and the results will be returned to the user.

Table 3. Characteristics of the benchmark query templates.

	q2	q8	q9	q10	q11	q12
Simple filters	✓	✓	✓		✓	✓
Complex filters				✓		
Number of fields selected	12	9	7	2	10	9
Number of joined tables	3	2	3	2	1	3
Join type	Star	Path	Star	Path		Path
Distinct				✓		
Order by		✓		✓		
Limit		✓		✓		

4 Implementation and Evaluation

All algorithms reported in this paper were implemented as a Java API named *CassandraJoins*. The API is going to be released, under an open source license, as soon as our results are published. Similarly to other JDBC drivers, using our API, users are able to call the method *CassandraJoins.connect* to get a new Cassandra cluster connection. Then they can call the *CassandraJoins.executeQuery* method to execute specific CQL queries containing joins. Finally, *CassandraJoins.close* is used to close the connection with the database. We have to note that the users *don't have to do any other configuration in order for the driver to work.*

Set-up. To perform our evaluation we used DataStax Enterprise 5.0 installed on a cluster running Ubuntu 15.04 LTS, with 4 GB of main memory each and a 4 CPU running at 3.0 GHz. The execution time reported in each case is the average of 50 runs of each query execution.

Dataset. To experiment with query execution times we used (a) a real dataset collected from 100 patients participating in iManageCancer [12] clinical trials (and the queries used for transforming those into triples). The queries used include 10 queries joining the two major tables available in the database requesting various fields with various conditions, (b) the standard BSBM benchmark [24] for NoSQL databases for RDF processing. This benchmark model is built around an e-commerce use case, in which different vendors offer a set of products, and consumers have posted reviews about products. The benchmark query mix included, illustrates the search and navigation pattern of a consumer looking for a product. As some of the generated queries are not supported by Cassandra (6 out of 12 templates queries were using nested selects, *like* and *in/not* operators), we omitted those from our queries. We have to note that based on these template queries, through the query generator provided, we created one hundred queries with different parameters. The characteristics of the query templates used, are shown in Table 3. This benchmark focuses on more common queries that appear when transforming data to triples, which usually requires a limited amount of joins as identified in this benchmark. Since the purpose of the *query optimization and execution module* in our architecture, is to enable the transformation of information to be further queried using SPARQL it is ideal for using it in our case.

Competitors. To get an indicative view on the efficiency of our implementation we compared our approach with the Simba-DataStax ODBC 2.4 driver.

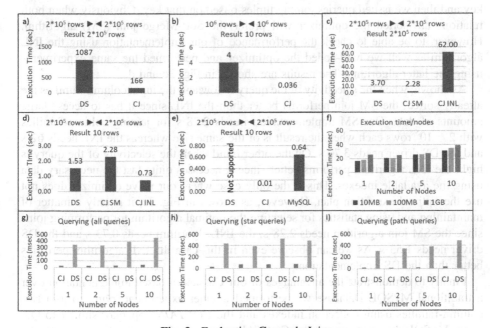

Fig. 3. Evaluating CassandraJoins

Real data benchmark. Initially, we performed real-world experiments using real data from the iManageCancer EU project and the queries we used to transform data to triples. Data included health records from the 100 *patients* participated within the project with written informed consent. As the volume of data eventually collected from those 100 patients was limited, in the real data benchmark we focus on a single node, eliminating also the multi-node network transfer overhead.

The first series of experiments uses queries joining the two major tables of the database with a join on the indexed field each time. Since we have indexes on the joined field the *CassandraJoins* (CJ) optimizer is using the Index-Nested Loops

(INL) join algorithm whereas we cannot identify the specific algorithm used by Simba-DataStax (DS) the source code is not publicly available. The results are shown in Fig. 3a and b for different input and output sizes.

We can observe that CJ is by far more efficient than the DS implementation in all cases by at least one order of magnitude. For example, when joining column-families with $2 * 10^5$ rows each and the result is of the same size our approach needs 166 s whereas the DS driver needs 1087 s as shown in Fig. 3a. Similarly, when joining column-families with 10^6 tuples each and the result is only 10 tuples our implementation needs 0,036 s whereas the DS driver needs 4 s as shown in Fig. 3b. Obviously, when the selectivity of the query is increased (less tuples are returned at the result) the execution time is decreased. This is reasonable since Cassandra is known to be extremely slow when a query needs to retrieves all rows of a table, whereas it is extremely fast when only a small subset of the rows is selected.

In the next set of experiments performed we ordered the data according to an index key and then we issued queries joining tuples based on that key. Obviously when both relations are sorted the optimizer will prefer the Sort-Merge join (SM) algorithm. However, to be able to check the performance of our implementation using the INL algorithm as well we hard-coded the optimizer and performed the same experiments using the latter algorithm. The results are shown in Fig. 3c and d.

As shown, when the selectivity of the query is low and the input column-families are already sorted, the SM join performs better than the INL since it has to access large amounts of data. Our SM implementation needs 2,28 s for joining column-families with $2 * 10^5$ rows each when the result is of the same size whereas the INL needs 62 s and the DS driver needs 3,70 s. On the other hand when the selectivity of the query is high, the INL algorithm performs better since it accesses quite fast the necessary data using the available indexes. This is the only case where our naïve optimizer does not use the optimal execution plan. However, as shown in Fig. 3c, it is only a matter of maintaining and using statistics for selecting the optimal algorithm for performing joins since the SM join algorithm needs 2,28 s, the INL algorithm needs 0,73 s and the DS driver needs 1,53 s. As we can observe even in this case the INL algorithm performs better than the DS implementation.

Finally to demonstrate the advantages of our Cassandra-based implementation over a traditional MySQL Database we performed another experiment trying to join two column-families using collection indexes and the *CONTAINS* operator similar to Q3. Since MySQL does not support the *CONTAINS* operator for sets we had to model the database there using 3 tables –one table for storing the set values. Then the corresponding query for MYSQL has to include three joins to retrieve the same results as the query issued in Cassandra. The results are shown in Fig. 3e. DS does not support joins on collections so we could not execute this query using that driver. We can observe that using CJ we need 0,01 s whereas using MySQL we need 0,64 s so the combined approach CassandraJoins-Cassandra is better than MySQL for joining collections.

BSBM benchmark. The next set of experiments elaborates on the average execution time for the BSBM benchmark queries over various node configurations using CJ. The results are shown in Fig. 3f. As shown, the more computational nodes involved, the higher the execution time. This is reasonable as communication cost increases as the number of nodes increases as well. In addition, as expected, the bigger the size of data the higher the execution time of the queries which is also reasonable.

The next set of experiments focus on comparing CJ and DS implementation. The results are shown for all queries in Fig. 3g for 1 GB of data for various node configuration and the time reported is the average execution time for all benchmark queries. As shown, CJ strictly dominates the Simba driver implementation giving one order of magnitude better execution times. In addition, as we can see, as the number of nodes increases, again the added communication cost has an impact on the overall execution time of the benchmark queries.

Focusing only on star queries (Fig. 3h) or only on path queries (Fig. 3i) still CJ is better than DS by one order of magnitude in all cases. Path queries seem to be more efficient than star queries in addition. Regarding individual queries from our results we identify that query templates q2 and q8 have the biggest execution time among all queries as a high number of tuples are returned to the user (q2).

5 Conclusions and Discussion

In this paper, we present an implementation of a query execution and optimization module on top of Cassandra databases that *does not impose any configuration overhead and can directly be used* by end-users in a plug-and-play manner. This module is consisted of three individual components: the *rewriter* that parses the input CQL queries, the *planner* that tries to comply with Cassandra restrictions and selects optimal join algorithms using simple heuristics and the *combiner* that executes the join and produces the final results. Two join algorithms were implemented, i.e. the *Index-Nested Loops* and the *Sort-Merge* join algorithms, extended to handle a new type of indexes available, the collection indexes. Our experiments demonstrate the advantages of our solution and confirm that our algorithms run efficiently and effectively. Our approach strictly dominates the commercial Simba-DataStax Driver currently available, in most cases by one order of magnitude. In addition, our implementation supports joins over collection indexes, not currently supported by any available implementation. Our implementation is freely available and will be released soon as open source. As such, it is currently the only available non-commercial implementation implementing joins over Cassandra databases.

However, deployment universality and simplicity comes at a cost. Making the driver oblivious of the underlying database statistics, data placement etc. misses big opportunities for optimization based on statistics and implementation of parallel algorithms for joining data. Not allowing infrastructure updates (e.g. installation of additional software at the cluster nodes) contributes more to this effect. As such, we are aware that often, the proposed approach has to retrieve a large volume of data that needs to be transferred through the network (in a distributed setup) and joined in a single end-user machine. This might lead to significant transfer cost and computational overhead. Nevertheless using our approach end users can directly benefit from both Cassandra's linear scalability for querying individual column families and use joins when needed with no additional set-up cost, with a small execution overhead, as our experiments show.

In addition, our framework opens up the possibilities for further enhancements gradually investigating how (a) we could unobtrusively retrieve statistics about the underlying data distribution and placement in order to exploit them for query optimization and (b) how the installation of additional modules in the available infrastructure will open up possibilities for larger parallelization of the computations. Eventually, as future work we envisage the integration of our algorithms directly in the Cassandra CQL language instead of placing a module on top of the CQL compiler. Adding joins directly to the CQL compiler will allow join operations to run faster than running them through an added layer and will make it easier for more joining algorithms common to RDBMSs to be implemented, (semi-joins or hash-joins for example). Cassandra is already doing hashing on the partition keys to find the rows, so Hash Join is the next algorithm to be implemented in our solution.

Without a doubt NoSQL databases is an emerging area of data managements that will only become more critical as the amount of data available explodes.

Acknowledgements. This work was partially supported by the EU project Bounce (H2020, #777167).

References

1. Chang, F., et al.: Bigtable: a distributed storage system for structured data. In: Seventh Symposium on Operating System Design and Implementation, pp. 205–218 (2006)
2. Sivasubramanian, S.: Amazon dynamoDB: a seamlessly scalable non-relational database service. In: ACM SIGMOD Conference, pp. 729–730 (2012)
3. Kim, W., Jeong, O.R., Kim, C.: A holistic view of big data. Int. J. Data Warehous. Min. (IJDWM) **10**(3), 59–69 (2014)
4. Madaan, A., Chu, W., Daigo, Y., Bhalla, S.: Quasi-relational query language interface for persistent standardized EHRs: using NoSQL databases. In: Madaan, A., Kikuchi, S., Bhalla, S. (eds.) DNIS 2013. LNCS, vol. 7813, pp. 182–196. Springer, Heidelberg (2013). https://doi.org/10.1007/978-3-642-37134-9_15
5. Yang, C.T., Liu, J.C., Hsu, W.H., Lu, H.W., Chu, W.C.C.: Implementation of data transform method into NoSQL database for healthcare data. In: PDCAT, pp. 198–205 (2013)
6. Ntarmos, N., Patlakas, I., Triantafillou, P.: Rank join queries in NoSQL databases. Proc. VLDB Endow. (PVLDB) **7**(7), 493–504 (2014)
7. Kim, C., Shim, K.: Supporting set-valued joins in NoSQL using MapReduce. Inform. Syst. J. **49**, 52–64 (2015)
8. Tatemura, J., Po, O., Hsiung, W.P., Hacigümüs, H.: Partiqle: an elastic SQL engine over key-value stores. In: SIGMOD Conference, pp. 629–632 (2012)
9. Google App Engine. https://cloud.google.com/appengine/. Accessed May 2019
10. Baker, J., et al.: Megastore: providing scalable, highly available storage for interactive services. In: CIDR, pp. 223-234 (2011)
11. Srinivasan, M., Gutkind, E.: How to do Joins in Apache Cassandra and DataStax Enterprise, DatStax Blog (2015). http://www.datastax.com/2015/03/how-to-do-joins-in-apache-cassandra-and-datastax-enterprise. Accessed May 2019
12. Kondylakis, H., Bucur, A., Dong, F., et al.: iManageCancer: developing a platform for empowering patients and strengthening self-management in cancer diseases. In: IEEE CBMS, pp. 755–760 (2018)

13. Marketakis, Y., Minadakis, N., Kondylakis, H., et al.: X3ML mapping framework for information integration in cultural heritage and beyond. IJDL J. **18**(4), 301–319 (2017)
14. Minadakis, N., Marketakis, Y., Kondylakis, H., et al.: X3ML framework: an effective suite for supporting data mappings. In: EMF-CRM@TPDL, pp. 1–12 (2015)
15. Kondylakis, H., Koumakis, L., Tsiknakis, M., Marias, K.: Implementing a data management infrastructure for big healthcare data. In: BHI, pp. 361–364 (2018)
16. Kondylakis, H., Koumakis, L., Katehakis, D., et al.: Developing a data infrastructure for enabling breast cancer women to BOUNCE back. In: IEEE CBMS (2019, to appear)
17. Kondylakis, H., Fountouris, A., Plexousakis, D.: Efficient implementation of joins over cassandra DBs. In: EDBT, pp. 666–667 (2016)
18. http://en.wikipedia.org/wiki/NoSQL. Accessed May 2019
19. Cattell, R.: Scalable SQL and NoSQL data stores. SIGMOD Rec. **33**(4), 12–27 (2010)
20. Indrawan-Santiago, M.: Database research: are we at a crossroad? Reflection on NoSQL. In: 15th BNiS Conference, pp. 45–51 (2012)
21. Hewitt, E.: Cassandra - The Definitive Guide: Distributed Data at Web Scale. Springer, Heidelberg (2011). ISBN 978-1-449-39041-9, pp. I-XXIII, 1-301
22. Aggarwal, A., Vitter, J.S.: The input/output complexity of sorting and related problems. Commun. ACM **31**(9), 1116–1127 (1988)
23. Ramakrishnan, R., Gehrke, J.: Database Management Systems. 2nd edn. McGraw-Hill Higher Education, New York
24. Bizer, C., Schultz, A.: The Berlin SPARQL benchmark. Int. J. Semant. Web Inf. Syst. **5**(2), 1–24 (2009)

MapReduce Join Across Geo-Distributed Data Centers

Giuseppe Di Modica and Orazio Tomarchio[✉]

Department of Electrical, Electronic and Computer Engineering,
University of Catania, V.le A. Doria 6, 95125 Catania, Italy
{Giuseppe.DiModica,Orazio.Tomarchio}@dieei.unict.it

Abstract. MapReduce is with no doubt the parallel computation paradigm which has managed to interpret and serve at best the need, expressed in any field, of running fast and accurate analyses on Big Data. The strength of MapReduce is its capability of exploiting the computing power of a cluster of resources, by distributing the load on multiple computing units, and of scaling with the number of computing units. Today many data analysis algorithms are available in the MapReduce form: Data Sorting, Data Indexing, Word Counting, Relations Joining to name just a few. These algorithms have been observed to work fine in computing context where the computing units (nodes) connect by way of high performing network links (in the order of Gigabits per second). Unfortunately, when it comes to run MapReduce on nodes that are geographically distant to each other the performance dramatically degrades. Basically, in such scenarios the cost for moving data among nodes connected via geographic links counterbalances the benefit of parallelization. In this paper the issues of running MapReduce Joins in a geo-distributed computing context are discussed. Furthermore, we propose to boost the performance of the Join algorithm by leveraging a hierarchical computing approach.

Keywords: MapReduce · Hadoop · Geo-distributed computation ·
Join · Hierarchical MapReduce

1 Introduction

The extraction of insightful information from distributed huge amount of data has become a fundamental part of all effective decision making processes. There are several different scenarios where having a framework able to elaborate geo-distributed data is fundamental. For example many organizations operate in different countries: data generated by customers' transactions are stored in different datacenters across the globe. Data can be distributed across different locations even in the same country. Organizations may decide to use multiple public or private clouds to increase reliability and security, without wishing to impair performance.

Again, many IoT-based applications, such as climate monitoring and simulation, have to analyze huge volumes of data sensed in multiple geographic

M. Younas et al. (Eds.): Innovate-Data 2019, CCIS 1054, pp. 18–31, 2019.
https://doi.org/10.1007/978-3-030-27355-2_2

locations to provide their final output. Other kind of applications need to work with data produced in research laboratories all around the globe: let us think, for instance, to bioinformatic applications such DNA sequencing or molecular simulations. All these applications have in common the need to store and process large amounts of geo-distributed data.

For what concerns the storage, geo-distributed databases and systems have been in existence for a long time [17]. However, these systems are not highly scalable, flexible, good enough for massively parallel processing, simple to program, and fast in answering a query. From the computing perspective, we do have robust and effective technology to mine big data. Parallel computing techniques such as the MapReduce [6] leverages the power provided by many computing resources (nodes) to boost the performance of the data analysis. Apache Hadoop [18] is the most prominent (open source) computing framework implementing the MapReduce paradigm. Unfortunately, Hadoop can guarantee a high performance only in computing contexts where computing resources are configured to form a cluster of homogeneous nodes, interconnected through high speed network links, and all the data are available within a single cluster, i.e., are not split into multiple data chunks which reside outside the cluster [9].

Several proposal have emerged in the scientific literature trying to enhance performance of MapReduce when dealing with geo-distributed data [7]. In a former work, we presented H2F [5], a hierarchical MapReduce based framework designed to elaborate big data distributed in geographical scenarios. In this paper, we enhance H2F in order to also deal with database query, by addressing the issues that arise in joining data that reside in distributed datacenters.

The rest of the paper is organized in the following way. Section 2 presents some background information and related work. In Sect. 3 we briefly recall the main design principles of the H2F framework. Then in Sect. 4 we describe how to deal with join issues by using our framework. Finally, we conclude our work in Sect. 5.

2 Background and Related Work

In this section we provide some background on standard join and their evolution on distributed databases. Then we will discuss existing proposals for joins and, more in general, database query in a mapreduce environment. Finally we review research works that address the processing of geo-distributed big data by using a MapReduce paradigm.

2.1 Background

When dealing with joins, it is common to distinguish between joins involving only two tables (*Two-Way Joins*) and joins involving more than two tables (*Multi-Way Joins*):

- *Two-way Joins*: Given two dataset P and Q, a two-way join is defined as a combination of tuples $p \in P$ and $q \in Q$, such that $p.a = q.b$. a and b are values from columns in P and Q respectively on which the join is to be done. Note that this is specifically an 'equi-join' in database terminology. This can be represented as: $P \bowtie_{a=b} Q$

– *Multi-way Joins*: Given n datasets $P_1, P_2, ..., P_n$, a multi-way join is defined as a combination of tuples $p_1 \in P_1, p_2 \in P_2, ..., p_n \in P_n$, such that $p_1.a_1 = p_2.a_2 = ... = p_n.a_n$. $a_1, a_2, ..., a_n$ are values from columns in $P_1, P_2, ..., P_n$ respectively on which the join is to be done. Notice once again that this is specifically an 'equi-join'. This can be represented as: $P \bowtie_{a=b} Q$

Joins have been widely studied and there are various algorithms available to carry out them, such as Nested-Loops Join, the Sort-Merge Join and the Hash Join [8]. All DBMSs support more than one algorithm to carry out joins.

The *Nested Loops Join* is the simplest and oldest join algorithms. It is composed of two nested loops that, respectively, analyze all the tuples looking for those who meet the join condition. It is capable of joining two datasets based on any join condition. However, the performance of this algorithm are very low when compared with the next two. The *Sort-Merge Join* sorts both datasets on the join attribute and then looks for tuples who meet the join condition by essentially merging the two datasets. The sorting step groups all tuples with the same value in the join column together and thus makes it easy to identify partitions or groups of tuples with the same value in the join column. The *Hash Join* algorithm consists of a 'build' phase and a 'probe' phase. The smaller dataset is loaded into an in-memory hash table in the build phase. Then, in the 'probe' phase, the larger dataset is scanned and joined with the relevant tuple(s) by looking into the hash table. This algorithm usually is faster than the Sort-Merge Join, but puts considerable load on memory for storing the hash-table.

Let us see now how join can be implemented when data (tables) are distributed over different nodes. In particular we briefly describe how join has been implemented in MapReduce environments and the issues that arise under these conditions: reduce-side join, map-side join and broadcast join. For the sake of simplicity, we describe them in their two-way version: however, they can be extended to the multi way case [19].

The *Reduce-Side Join* algorithm is in many ways a distributed version of Sort-Merge Join. In this algorithm, as the name suggests, the actual join happens on the Reduce side of the framework. The 'map' phase only pre-processes the tuples of the two datasets to organize them in terms of the join key. Then, each sorted partition is sent to a reduce function for merging.

The Reduce-Side join seems like the natural way to join datasets using Map/Reduce. It uses the framework's built-in capability to sort the intermediate key-value pairs before they reach the Reducer. But this sorting often is a very time consuming step. The *Map Side join* exploits another way of joining datasets. This functionality is present out of the box and is one of the fastest way to join two datasets using Map/Reduce, but places severe constraints on the datasets that can be used for the join: the sort ordering of the data in each dataset must be identical for datasets to be joined, a given key has to be in the same partition in each dataset so that all partitions that can hold a key are joined together and, a given key has to be in the same partition in each dataset so that all partitions that can hold a key are joined together. These constraints are quite strict but are all satisfied by any output dataset of a Hadoop job. Thus, a pre-processing step is needed, where we simply pass both the dataset through a basic Hadoop job which do no processing on the data but simply pass it through the framework which partitions, groups and sorts it. The output is compliant with all the constraints mentioned above. The overall gain of Map Side join is due to the potential for the elimination of the reduce phase and/or the great reduction in the amount of data required for the reduce.

The *Broadcast Join* algorithm is a distributed variant of the Hash Join algorithm. If one of the datasets is very small, such that it can fit in memory, then there is an optimization that can be exploited to avoid the data transfer overhead involved in transferring values from Mappers to Reducers. This kind of scenario is often seen in real-world applications [3]. This small dataset is sent to every node in the distributed cluster. It is then loaded into an in-memory hash table. Each map function streams through its allocated chunk of input dataset and probes this hash table for matching tuples. This algorithm is thus a Map-only algorithm.

2.2   Distributed Joins in MapReduce Environments

Starting from the previously described distributed joins several optimization have been proposed [1] for MapReduce environments. As already stated in the introduction, the main problem in this case is related to the optimization of the join operation when dealing with tables on different nodes of the cluster [16].

SquirrelJoin [15] a distributed join processing technique that uses lazy partitioning to adapt to transient network skew in clusters. It recognizes that redistributing a large number of data records is limited by the available network bandwidth between nodes: since it may accounts for a relevant percentage of the join completion time, it proposes a runtime mitigation technique for network skew. The work in [14] proposes Flow-Join, another specific distributed join processing technique. It has been designed for cluster whose node are interconnected by means of high speed network (Infiniband). Its basic idea is to detect heavy hitters at runtime using small approximate histograms and to adapt the redistribution scheme to resolve load imbalances before they impact the join performance. The work in [1] investigates algorithms to optimize the join operation in a map-reduce environment. In particular, it identifies specific cases where the proposed algorithm outperform the conventional way of using map-reduce to implement joins.

From the analysis of these previous works it emerges that none of these approaches is designed and takes into account a scenario where data are placed on geographically distributed data center. They have been designed and analyzed for working within a single datacenter, where nodes are generally connected through high speed network technologies.

When trying to apply the same algorithms to a geographically distributed datacenters, they exhibit high query response times because these frameworks cannot cope with the relatively low and variable capacity of WAN links. On the other hand, the naive approach to aggregate all the datasets to a central site (a large DC), before executing the queries, is not effective when dealing with huge amount of data (big data).

2.3   MapReduce in Geo-Distributed Environments

The works in literature that address the processing of geo-distributed big data by using a MapReduce paradigm can be roughly divided in two main approaches: (a) enhanced versions of the Hadoop framework which try to get awareness of the heterogeneity of computing nodes and network links (*Geo-hadoop* approach); (b) hierarchical frameworks which split the MapReduce job into many sub jobs that are sent to as many computing nodes hosting Hadoop instances which do the computation and send back the result to a coordinator node that is in charge of merging the sub jobs' output (*Hierarchical* approach).

The former approach's strategy is to optimize the Hadoop's native steps: Push, Map, Shuffle, Reduce. To reduce the job's average makespan, these four steps must be smartly coordinated. Some works have launched modified version of Hadoop capable of enhancing only a single step [11,13]. Heintz et al. [9] study the dynamics of the steps and claim the need for a comprehensive, end-to-end optimization of the job's execution flow. They propose an analytical model which looks at parameters such as the network links, the nodes capacity and the applications profile, and turns the makespan optimization problem into a linear programming problem solvable by means of Mixed Integer Programming techniques. In [22] authors propose a modified version of the Hadoop algorithm that improves the job performance in a cloud based multi-data center computing context. Improvements span the whole MapReduce process, and concern the capability of the system to predict the localization of MapReduce jobs and to prefetch the data allocated as input to the Map processes. Changes in the Hadoop algorithm regarded the modification of the job and task scheduler, as well as of the HDFS' data placement policy. In [2], authors propose an approach called Meta-MapReduce that decreases the communication cost significantly. It exploits the locality of data and mappers-reducers avoiding the movement of data that does not participate in the final output. To reach this goal it provides a way to compute output using metadata which are much smaller than the original data. Although not expressly designed for geo-distributed data, the same authors state that Meta-MapReduce can be extended also for geographically distributed data processing.

In [20] authors propose a MapReduce framework that aims to enable large-scale distributed computing across multiple High End Computing (HEC) clusters. The proposed framework's main innovative ideas are the introduction of a distributed file system manager (Gfarm) to replace the original Hadoop's HDFS and the employment of multiple task trackers (one per each cluster involved in the computation). The job execution flow enforced by the proposed framework basically resembles that of the Hadoop framework: tasks are assigned to those clusters where the required input data is present, disregarding both the computing power and the interconnecting links of the clusters involved in the computation. The tangible benefits are bounded to the bandwidth usage, for which G-Hadoop shows a better data traffic management.

On the other hand hierarchical approaches can be found: their common idea is to make a smart decomposition of the job into multiple sub jobs and then exploit the native potential of the plain Hadoop. They mainly envisions two computing levels: a *bottom level*, where plain MapReduce is run on locally available data, and a *top level*, where a centralized entity coordinates the activities of splitting the workload into many sub jobs, as well as gathering and packaging the sub jobs' output. For this kind of approach, a strategy must to be conceived to establish how to redistribute data among the available clusters in order to optimize the job's overall makespan.

In [12] authors introduce a hierarchical MapReduce architecture and a load-balancing algorithm that distributes the workload across multiple data centers. The balancing process is fed with the number of cores available at each data center, the number of Map tasks runnable at each data center and the features (CPU or I/O bound) of the job's application. The authors also suggest to compress data before migrating them among data canters. Jayalath et al. [10] list the issues connected to the execution of the MapReduce on geographically distributed data. They specifically address a scenario where multiple MapReduce operations need to be performed, one after the other, on the same data. They lean towards a hierarchical approach, and propose to represent all the possible jobs' execution paths by means of a data transformation graph to be used for the determination of optimized schedules for job sequences. The well-known Dijkstra's shortest path algorithm is then used to determine the optimized schedule. In [21] authors introduce an extra MapReduce phase named "merge", that works after map and reduce phases, and extends the MapReduce model for heterogeneous data. The model turns to be useful in the specific context of relational database, as it is capable of expressing relational algebra operators as well as of implementing several join algorithms.

3   Hierarchical Hadoop Framework

The Hierarchical Hadoop Framework (H2F) [4] is a framework designed to efficiently elaborate Big Data in highly distributed and geographical scenarios. The proposed approach revolves around the intuition that in a scenario where big data are sparsely distributed among multiple heterogeneous data centers connected via geographical network links, in order to efficiently run MapReduce computation on the data all available resources (computing capacity and network links) must be exploited at best. An job computation is as more efficient as shorter is time it takes to complete the job request (known as **job makespan**).

In principle, given a job that needs to run some analysis on distributed data, the strategy enforced by the H2F is to seek for data centers that have high and available computing power, move as much data as possible to those data centers by means of fast network links, let each data center run computation on the received data (by triggering a local MapReduce job), and finally gather and merge the output of such parallel and distributed computation. Apart from the data centers' computing power and network links' bandwidth, the H2F scheduling strategy takes in great consideration the "profile" of the application algorithm that is going to crunch the data. An application profile tells how fast the application is to produce data in output (*Throughput*) and the capacity of the application to compress/expand the amount of data received in input when producing the output. So, given a distributed computing scenario, the job scheduling of two applications showing different application profiles may look very different, as each application's specific "behaviour" will impact on the choices of the data to move, the network links to use for the data shift and the data centers to involve in the computation.

Figure 1 shows a basic reference scenario addressed by the H2F. Data centers (called *Sites*) populate the bottom level of a two-level computing hierarchy. Each Site owns a certain amount of data and is capable of running plain Hadoop jobs. Upon receiving a job request, a Site performs the whole MapReduce process on the local cluster(s) and returns the result of the elaboration to the top-level. The *Top-level Manager* owns the system's business logic and is in charge of the management of the geo-distributed parallel computing. Upon the submission of a top-level job, the business logic schedules the set of sub-jobs to be spread in the distributed context, collects the sub-job results and packages the overall output. In the depicted scenario the numbered arrows describe a typical execution flow triggered by the submission of a top-level job. This specific case envisioned a shift of data from one Site to another Site, and the run of local MapReduce sub-jobs on two Sites. Here follows a step-by-step description of the actions taken by the system to serve the job:

1. The Top-Level Manager receives a job submission.
2. A Top-level Job Execution Plan is generated (TJEP), using information about (a) the status of the bottom level layer like the distribution of the data set among Sites, (b) the current computing capabilities of Sites, (c) the topology of the network and (d) the current capacity of its links.
3. The Top-Level Manager executes the TJEP. Following the plan instructions, it orders Site1 to shift some data to Site4.
4. The actual data shift from Site1 to Site4 takes place.
5. According to the plan, the Top-Level Master sends a message to trigger the subjobs execution on the Sites where data are residing. In particular, top-level Map tasks are triggered to run on Site2 and Site4 respectively (we remind that a top-level Map task corresponds to a Hadoop subjob).
6. Site2 and Site4 executes local Hadoop subjobs on their respective data sets.
7. Sites sends the results obtained from local executions to the Top-Level Manager.
8. A Global Reducer routine within the Top-Level Manager collects all the partial results coming from the bottom level layer and performs the reduction on this data.
9. Final result is forwarded to the Job submitter.

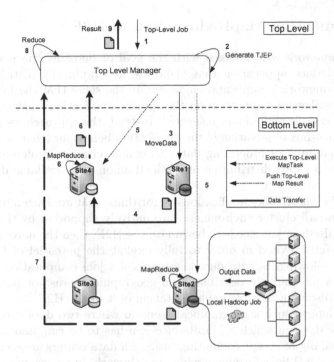

Fig. 1. Job Execution Flow

The whole job execution process is transparent to the submitter, who just needs to provide the job to execute and a pointer to the target data the job will have to process. When a job is submitted to the H2F, the job scheduler takes into account the application profile and dynamics of the availability of both the computing and the network resources to find the best combination of resources that guarantees the shortest makespan to the job. It first generates all the possible job execution paths in the addressed distributed computing context. Each generated path is then assigned a score, which is a proportional to the esti-mated job makespan (the shorter the estimated completion time, the higher the score). The path exhibiting the shortest makespan (best score) will be selected to enforce the job execution.

Finally, in Fig. 1, according to the scheduling directives, all the data natively residing in Site1 are shifted to Site4. The job scheduler might also decide, if deemed convenient, to fragment the data of a Site into smaller pieces (fragments) and shift fragments to multiple Sites available. The introduction of this feature increases the number of potential execution paths, but at the same time increases the chance to find a high performer path.

4 Implementing MapReduce Join in H2F

The H2F framework was designed with the goal of boosting the performance of any MapReduce application that elaborate geographically distributed Big Data. The framework is application-agnostic, in the sense that the job scheduling strategy makes no assumption on the specific application's pattern nor the application's code is preliminary inspected. Instead, the approach we propose is an off-line black-box observation of the application behaviour (that is, the application's "capability" of processing data) to characterize its profile and, based on that, make a smart distribution of the load among the available distributed clusters.

Most of the MapReduce applications' algorithms, that are known to be working in as standard cluster environment, are natively supported by the H2F [4]. But, if a MapReduce join needs to be run on the H2F, then the naive algorithm must be carefully revised in order to fully exploit the potential of the framework. In the following, we introduce a scenario of a join computation to be run on tables that are natively distributed on geographically distant data centers, and then we discuss a possible implementation of it in the H2F.

Figure 2 depicts such a geographical scenario where two data centers (clusters) hold the data on which a MapReduce join has to be run, and some other clusters are willing to accept computing tasks. All data centers are connected by way of geographical links forming a wide area network. In the example scenario, $Cluster_1$ holds the relation $U(A, B)$ while $Cluster_2$ holds the relation $V(B, C)$, being B the attribute on which a join has to be performed. The approach we propose takes inspiration from the one presented in [2] (called "Meta-MapReduce") and retains from that work the following assumption: the size of all the B values is very small if compared to the size of the values of the attributes A and C. Should that assumption not hold true, the Meta-Mapreduce would bring no benefit in terms of communication costs with respect to the original MapReduce join algorithm.

According to the Meta-MapReduce proposal, there are two off-line preprocessing steps to be taken on the original input data before running the real join job: (a) generating the data Index and (b) generating the Meta-data. In [2] those operations are given for granted by authors, whose main concern is optimizing the cost for moving data among geographically distant clusters. Yet, generating those data is a burden that can impact the whole job performance. H2F comes to the aid of the Meta-MapReduce join algorithm by performing a quick building of the two relations' indices and Meta-data. Basically, the idea is to exploit scheduling capability of the H2F to smartly fragment the two relations into small pieces (fragments), distribute them to the most appropriate clusters that guarantee for a fast, parallel build of Indices and Meta-data of fragments, gather the partial outcomes of the distributed computation and finally assemble the result. In Fig. 3(a) the first step of a computation flow that might potentially be triggered to build the mentioned data is showed. Following the H2F taxonomy, $Cluster_1$ and $Cluster_2$ will host the *Global Reducers* responsible for assembling the output of the hierarchical computation of the Indices and the

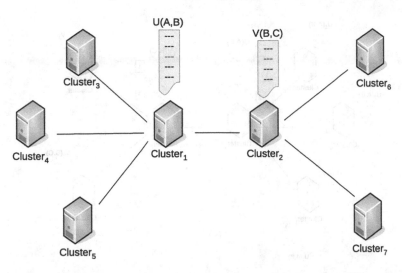

Fig. 2. Example scenario: MapReduce join over two relations natively located in two distant data centers

Meta-data of relations $U(A, B)$ and $V(B, C)$ respectively. $Cluster_3$, $Cluster_4$ and $Cluster_5$ have been selected by the H2F as *bottom-level nodes* on which to run the distributed computation of relation $U(A, B)$'s Index and Meta-Data, while $Cluster_6$ and $Cluster_7$ will do the same for relation $V(B, C)$'s data. In the figure, the reader may notice that the two relations have been fragmented into small pieces, and the fragments have been shifted to the selected working Clusters.

As mentioned in Sect. 3, the selection of the Clusters to be promoted as bottom-level nodes (working Clusters), as well as the determination of the size and the number of fragments that the relations have to be split into, is a task of the H2F's job scheduling algorithm. The one depicted in the Fig. 3(a) is just an example scenario reflecting the output of such an algorithm's execution.

The bottom-level computation starts right after the fragments have been delivered to the charged Clusters (see Fig. 3(b)). We remind that once a fragment reaches a Cluster, its elaboration is carried out by a vanilla Hadoop framework, that will transparently exploit all the Cluster's available nodes to run the job. Partial outputs (PO in the figures) elaborated by the Hadoop frameworks of all the working Clusters are then gathered and assembled by the Global Reducers residing in $Cluster_1$ and $Cluster_2$ (see Fig. 3(c), the overall output being a merge of the partial Indices and Meta-data elaborated by the working Clusters.

$Cluster_1$ and $Cluster_2$ now have the data required to start the Meta-MapReduce join computation. Following the example scenario, the join computation will begin with the Map join phase (see Fig. 4(a)), where selected Mappers are responsible for building the intermediate output out of the Meta-Data provided as input. According to [2], the Mappers will produce intermediate outputs of the form $<b_i, values(b_i)>$, being b_i a joining key and $values(b_i)$ the size of

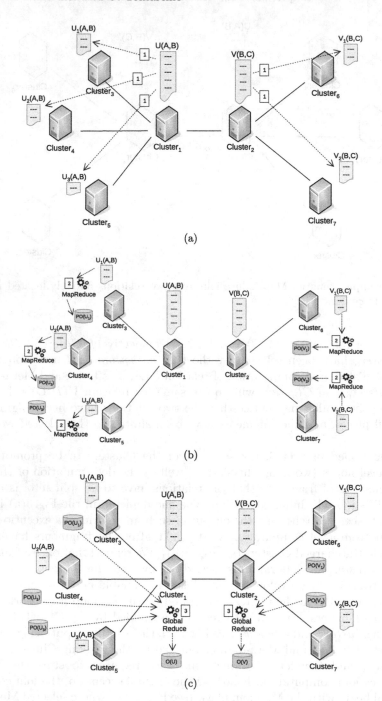

Fig. 3. Building Indices and Meta-Data exploiting the hierarchical computation of the H2F: (a) data fragmentation and distribution; (b) building of partial output; (c) data gathering and global reducing

the non-joining attributes associated with b_i. Join Mappers can be scheduled to run right on the Clusters where partial Indices and Meta-Data have been built, i.e., on $Cluster_3$, $Cluster_4$ and $Cluster_5$ for relation $U(A, B)$, and on $Cluster_6$ and $Cluster_7$ for relation $V(B, C)$. Intermediate output (IO) elaborated by such Mappers are then the sent to Reducers which, given the distribution of the relations depicted in the Fig. 2, can be conveniently scheduled to run on $Cluster_1$ and $Cluster_2$. Here, Join Reducers receive the intermediate output, perform "call" operations of original data for all the b_i keys that have a correspondent value in both the relations and eventually make the join of original data (see Fig. 4(b)). By having placed the Join Reducers on $Cluster_1$ and $Cluster_2$ respectively, the number of remote "call" operations is kept low. For instance, the Reducer residing on $Cluster_1$ will make local calls (LC) to retrieve original data from the relation $U(A, B)$, while remote calls (RC) are necessary to retrieve original data from relation $V(B, C)$; similarly the Reducer on $Cluster_2$ will make local calls to retrieve data from relation $V(B, C)$ and remote calls to retrieve data from $U(A, B)$.

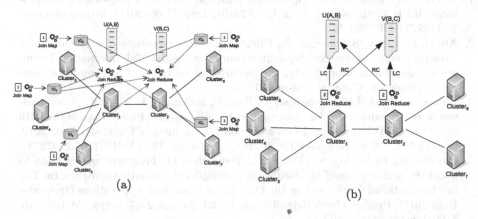

Fig. 4. Meta-MapReduce join: (a) Map step; (b) Reduce step

The proposed approach can be generalized for more complex join schemes such as the Multi-way join. Let us assume that we have three clusters, say $Cluster_1$, $Cluster_2$ and $Cluster_3$. $Cluster_1$ has two relations $U(A, B)$ and $V(B, C)$, $Cluster_2$ has two relations $W(D, B)$ and $X(B, E)$, and $Cluster_3$ has two relations $Y(F, B)$ and $Z(B, G)$, being B the attribute against which to run the Multi-way join. Each Cluster will independently run a hierarchical MapReduce routine to build indices and Meta-Data of the relations it owns. Once those data are available, the Meta-MapReduce join can begin. As better explained in [2], this job will consist in joining data between two clusters, and then joining the output with the data of the third cluster.

5 Conclusion

The big data era has called for innovative computing paradigm to handle huge amounts of data in a scalable and effective way. The MapReduce has best interpreted this need. Today the majority of the most used data analysis algorithms are available in the MapReduce form, i.e., can exploit the capacity of a cluster of multiple computing nodes. Unfortunately, when the computing power is geographically distributed, the efficiency of the MapReduce, as well as that of the data analysis' algorithms, degrade. In this paper an analysis of the Join algorithm and its MapReduce implementations was conducted. Further, an efficient way of implementing the MapReduce Join in a geographically distributed computing scenario was discussed. In the future, the benefit of the proposed approach will be assessed in a concrete test bed.

References

1. Afrati, F.N., Ullman, J.D.: Optimizing multiway joins in a map-reduce environment. IEEE Trans. Knowl. Data Eng. **23**(9), 1282–1298 (2011). https://doi.org/10.1109/TKDE.2011.47
2. Afrati, F., Dolev, S., Sharma, S., Ullman, J.: Meta-MapReduce: a technique for reducing communication in MapReduce computations. In: 17th International Symposium on Stabilization, Safety, and Security of Distributed Systems (Springer-SSS), Edmonton, Canada, August 2015
3. Blanas, S., Patel, J.M., Ercegovac, V., Rao, J., Shekita, E.J., Tian, Y.: A comparison of join algorithms for log processing in MapReduce. In: Proceedings of the 2010 ACM SIGMOD International Conference on Management of Data, SIGMOD 2010, pp. 975–986. ACM, New York (2010). https://doi.org/10.1145/1807167.1807273
4. Cavallo, M., Di Modica, G., Polito, C., Tomarchio, O.: Fragmenting Big Data to boost the performance of MapReduce in geographical computing contexts. In: The 3rd International Conference on Big Data Innovations and Applications (Innovate-Data 2017), Prague, Czech Republic, pp. 17–24, August 2017. https://doi.org/10.1109/Innovate-Data.2017.12
5. Cavallo, M., Modica, G.D., Polito, C., Tomarchio, O.: A hierarchical Hadoop framework to handle Big Data in geo-distributed computing environments. Int. J. Inf. Technol. Syst. Approach (IJITSA) **11**(1), 16–47 (2018). https://doi.org/10.4018/IJITSA.2018010102
6. Dean, J., Ghemawat, S.: MapReduce: simplified data processing on large clusters. In: Proceedings of the 6th Conference on Symposium on Operating Systems Design and Implementation, OSDI 2004. USENIX Association (2004)
7. Dolev, S., Florissi, P., Gudes, E., Sharma, S., Singer, I.: A survey on geographically distributed big-data processing using mapreduce. IEEE Trans. Big Data **5**(1), 60–80 (2019). https://doi.org/10.1109/TBDATA.2017.2723473
8. Garcia-Molina, H., Ullman, J.D., Widom, J.: Database Systems: The Complete Book, 2nd edn. Prentice Hall Press, Upper Saddle River (2008)
9. Heintz, B., Chandra, A., Sitaraman, R., Weissman, J.: End-to-end optimization for geo-distributed MapReduce. IEEE Trans. Cloud Comput. **4**(3), 293–306 (2016). https://doi.org/10.1109/TCC.2014.2355225

10. Jayalath, C., Stephen, J., Eugster, P.: From the cloud to the atmosphere: running MapReduce across data centers. IEEE Trans. Comput. **63**(1), 74–87 (2014). https://doi.org/10.1109/TC.2013.121
11. Kim, S., Won, J., Han, H., Eom, H., Yeom, H.Y.: Improving Hadoop performance in intercloud environments. SIGMETRICS Perform. Eval. Rev. **39**(3), 107–109 (2011). https://doi.org/10.1145/2160803.2160873
12. Luo, Y., Guo, Z., Sun, Y., Plale, B., Qiu, J., Li, W.W.: A hierarchical framework for cross-domain MapReduce execution. In: Proceedings of the Second International Workshop on Emerging Computational Methods for the Life Sciences, ECMLS 2011, pp. 15–22 (2011). https://doi.org/10.1145/1996023.1996026
13. Mattess, M., Calheiros, R.N., Buyya, R.: Scaling MapReduce applications across hybrid clouds to meet soft deadlines. In: Proceedings of the 2013 IEEE 27th International Conference on Advanced Information Networking and Applications, AINA 2013, pp. 629–636 (2013). https://doi.org/10.1109/AINA.2013.51
14. Rödiger, W., Idicula, S., Kemper, A., Neumann, T.: Flow-join: adaptive skew handling for distributed joins over high-speed networks. In: 2016 IEEE 32nd International Conference on Data Engineering (ICDE), pp. 1194–1205, May 2016. https://doi.org/10.1109/ICDE.2016.7498324
15. Rupprecht, L., Culhane, W., Pietzuch, P.: SquirrelJoin: network-aware distributed join processing with lazy partitioning. Proc. VLDB Endowment **10**(11), 1250–1261 (2017). https://doi.org/10.14778/3137628.3137636
16. Sarma, A.D., Afrati, F.N., Salihoglu, S., Ullman, J.D.: Upper and lower bounds on the cost of a map-reduce computation. Proc. VLDB Endowment **6**(4), 277–288 (2013). https://doi.org/10.14778/2535570.2488334
17. Sheth, A.P., Larson, J.A.: Federated database systems for managing distributed, heterogeneous, and autonomous databases. ACM Comput. Surv. **22**(3), 183–236 (1990). https://doi.org/10.1145/96602.96604
18. The Apache Software Foundation: The Apache Hadoop project (2011). http://hadoop.apache.org/
19. Venner, J.: Pro Hadoop, 1st edn. Apress, Berkely (2009)
20. Wang, L., et al.: G-hadoop: MapReduce across distributed data centers for data-intensive computing. Future Gener. Comput. Syst. **29**(3), 739–750 (2013). https://doi.org/10.1016/j.future.2012.09.001. Special Section: Recent Developments in High Performance Computing and Security
21. Yang, H., Dasdan, A., Hsiao, R., Parker, D.S.: Map-reduce-merge: simplified relational data processing on large clusters. In: Proceedings of the 2007 ACM SIGMOD International Conference on Management of Data, SIGMOD 2007, pp. 1029–1040 (2007). https://doi.org/10.1145/1247480.1247602
22. Zhang, Q., et al.: Improving Hadoop service provisioning in a geographically distributed cloud. In: 2014 IEEE 7th International Conference on Cloud Computing (CLOUD 2014), pp. 432–439, June 2014. https://doi.org/10.1109/CLOUD.2014.65

Mobile Device Identification via User Behavior Analysis

Kadriye Dogan and Ozlem Durmaz Incel$^{(\boxtimes)}$ (iD)

Department of Computer Engineering, Galatasaray University,
Ortakoy, 34349 Istanbul, Turkey
`kadriye.dogan@ogr.gsu.edu.tr`, `odincel@gsu.edu.tr`

Abstract. Modern mobile devices are capable of sensing a large variety of changes, ranging from users' motions to environmental conditions. Context-aware applications utilize the sensing capability of these devices for various purposes, such as human activity recognition, health coaching or advertising, etc. Identifying devices and authenticating unique users is another application area where mobile device sensors can be utilized to ensure more intelligent, robust and reliable systems. Traditional systems use cookies, hardware or software fingerprinting to identify a user but due to privacy and security vulnerabilities, none of these methods propose a permanent solution, thus sensor fingerprinting not only identifies devices but also makes it possible to create non-erasable fingerprints.

In this work, we focus on distinguishing devices via mobile device sensors. To this end, a large dataset, larger than 25 GB, which consists of accelerometer and gyroscope sensor data from 21 distinct devices is utilized. We employ different classification methods on extracted 40 features based on various time windows from mobile sensors. Namely, we use random forest, gradient boosting machine, and generalized linear model classifiers. In conclusion, we obtain the highest accuracy as 97% from various experiments in identifying 21 devices using gradient boosting machine on the data from accelerometer and gyroscope sensors.

Keywords: Sensor fingerprinting · User identification · Mobile device identification · Mobile device sensors

1 Introduction

Modern web or mobile applications aim to identify users uniquely to recommend more intelligent contents, products or display them only relevant ads. In addition to intelligent guidance, particular kind of applications need to remember the last state of a user to continue recent activity, such as showing shopping cart with items added before on an e-commerce system or starting a TV show from the last scene. Accordingly, traditional user identification systems use well known methodologies such as cookies, hardware and software fingerprinting. However, these solutions exhibit various disadvantages.

The first identifier, a cookie, can be removed from a browser by a user at any time and 3rd party applications can also access the information stored by a

© Springer Nature Switzerland AG 2019
M. Younas et al. (Eds.): Innovate-Data 2019, CCIS 1054, pp. 32–46, 2019.
https://doi.org/10.1007/978-3-030-27355-2_3

cookie. In addition to these problems, many security vulnerabilities are detected in browsers related to confidential data access such as credit card, email or passwords. Another identifier, hardware fingerprinting, represents device specific identification number and enables tracking users via device id such as IMEI (available only phones), WLAN MAC address, bluetooth MAC address, etc. However, applications which intend to use hardware fingerprint need to obtain user permission due to privacy regulations. Besides a user can change the device at anytime, so that the system loses one of the identified users.

The last identifier, software fingerprinting, consists of two main categories; browser and mobile advertisement fingerprinting. The browser fingerprinting methodology collects detailed information from a browser and the operating system, such as language, timezone, monitor settings, etc., and produces unique identification key for each user but in some cases this key can be similar to another individual due to the same configurations. For instance, two colleagues work in the same department of a bank and IT department configures all the notebooks exactly the same to reduce operation costs and then restricts specifications, such as language, timezone or plugin installations. Hence, the same browser fingerprint key will be generated for all users of the bank. Lastly, mobile advertisement fingerprinting is both enabled for Android (AAID) and IOS (IDFA) operating systems that allow access to user id for advertising for various marketing purposes. However, a user can reset his/her device at anytime or remove permission from an application so this key is only reliable as much as hardware fingerprinting.

In consideration of traditional fingerprinting problems, researchers proposed a new methodology, sensor fingerprinting, which detects behavioral patterns of a user via collected sensor data from mobile devices. Sensor fingerprinting solves both user permission and persistent identifier problems and makes easier to develop reliable and robust systems. Modern browsers are capable of accessing device sensors such as accelerometer, gyroscope, etc. without any permission so it is possible to collect and analyze data on the device in the background. Furthermore, sensor fingerprinting only needs sensor data rather than additional device identifiers such as IMEI, AAID, etc. so software or hardware changes does not affect user identification. Eventually, users carry their smartphones or smartwatches almost all the day and have routines in their interaction patterns on the devices, such as reading news on an application after a morning walk. Hence, applications can track user behaviours via mobile sensors effortlessly and distinguish users from each other with different routines.

In this paper, we explore the performance of sensor fingerprinting in order to identify mobile devices with a large dataset which is provided by CrowdSignals.io [3] and collected from different individuals via various devices. The dataset differs from others which are used in previous works [12] due to the diversity from various aspects, such as participant characteristics, data size, sensors types, etc. CrowdSignals.io aims to build the largest and richest mobile sensor dataset so researchers, students or developers can access labeled sensor data easily and focus on their researches. Moreover, the dataset represents not only sensor data but also additional information per participant, such as age range, sex, occupation, etc., thereby reliable machine learning models can be created to develop real-time applications.

In particular, we focus on the use of motion sensors, namely accelerometer and gyroscope, in identifying 21 devices/users, considering different classifiers and feature sets. Our aim is to explore recognition with only motion sensors and simple features which are not computationally expensive. We also explore different window sizes in the process, since our ultimate aim is to develop a real-time system working on the device. The highlights and contributions can be summarized as follows:

- Using only accelerometer sensor is more effective than gyroscope to identify devices correctly. However, combining sensors increases the performance of models considerably, around 6%. On the other hand, accelerometer can be used standalone when resource usage is a concern with multiple sensors.
- Our experimental results show that Gradient Boosting Machine is the most efficient algorithm among others since each trained tree helps the next one to minimize the errors. Furthermore, Random Forest provides a significant performance slightly less than Gradient Boosting Machine. However, Generalized Linear Model is not a proper classifier when compared to others providing lowest accurate results.

Rest of the paper is organized as follows: In Sect. 2, we present the related work and how our method differs from the related studies. Section 3 presents our methodology particularly the parameters and experiments considered in this study. In Sect. 4, we present the results of the experiments and discuss our findings. Finally, Sect. 5 concludes the paper and includes the future studies.

2   Related Work

Many studies have been presented in the area of sensor fingerprinting for different purposes, such as tracking users across applications, strong authentication or secure peer to peer data sharing. Existing studies show that accelerometer sensor has a great impact on identifying devices, hence it is commonly used in almost all experiments. In a previous study [11], authors explain the details of why accelerometer sensor is chosen as a basis for sensor fingerprinting: accelerometer chips produce different responses for the same motion even if the same chip model is used due to hardware faultiness during the manufacturing process. Therefore, devices can be easily differentiated. In that study, 80 accelerometer chips, 25 android tablets and 2 tablets are used and 96% precision and recall values are achieved with a bagged decision tree classifier.

In another study [10] it is shown that using accelerometer and gyroscope sensors together is an effective method to gain better performance from machine learning models and experiments demonstrate that using both sensors increases average F-score from 85%–90% range to 96%. In this study, we also observe similar findings and we show that combining accelerometer and gyroscope increases identification accuracies by 6% for different classifiers.

Table 1. CrowdSignals.io dataset fields

Field name	Description
x	A list of x axis values
y	A list of y axis values
z	A list of z axis values
user id	Unique identifier of user
timestamps	A list of timestamp values
type	Type of sensor

Sensor fingerprinting not only solves tracking user problems but also provides smarter solutions for secure authentication systems as discussed in [13]. In the study, authors reported accuracy as 98.1% which is well enough to distinguish device owner from other users who attempt to access the device. During experiments, authors used smartwatch and smartphone sensors together to improve accuracy and eliminated magnetometer and light sensors since they are affected from environmental conditions.

In another study [14], only accelerometer sensor is preferred to generate secret keys in order to propose new method for pairing devices and to provide device-to-device authentication. In another study [9], researchers aim to implement a new device identification system which is based on sensor fingerprinting via accelerometer and speakerphone-microphone sensors on a mobile device. However, authors report that only 15.1% of devices were identified correctly since the running code which collects accelerometer sensor data on mobile browser is unverified and not reliable.

Considering previous works, we aim to analyze the performance of complex models when compared to simple ones in this study. Therefore, we utilized Gradient Boosting Machine and Random Forest as complex models and compared the performance results with the Generalized Linear Model.

3   Methodology

In this section, we describe the steps of our methodology which are frequently used in machine learning studies. These steps consist of data collection, preprocessing, feature engineering, model training and evaluation.

3.1   Data Collection

One of the challenges faced in machine learning studies is the difficulty of collecting labelled datasets. A great effort is spent to find participants and develop data collection applications. Moreover, use of different datasets in different studies makes it difficult to generalize and compare the results. In this respect, publicly available and labelled datasets with high quality are important for reproducibility in the research community. CrowdSignals.io [3] provides a mobile sensor dataset [2,15] from various devices and makes it easier to access labelled data.

In this work, we used the CrowdSignals.io dataset which includes data collected from several sensors, such as accelerometer, gyroscope, magnetometer, etc., from 30 different Android users. The data was collected for more than 15 days and contains 5 GB of sensor records per user on average. In spite of having more than 21 users, we eliminated others due to lack of data from motion sensors. Furthermore, although the dataset includes other sensors, such as light, magnetometer, we only used accelerometer and gyroscope sensors since our aim is to explore the use motion sensors in the identification process, in this work. However, we plan to include other sensors in future work. Table 1 shows the utilized fields of the dataset during experiments.

In Table 2, details about the participants are provided. All users are Android users, 7 of them are female, age span is between 18 to 60. In addition to the participant characteristics, we also added the data size (sixth column) per user.

3.2   Sensor Types

In this section, we describe how we used the data from accelerometer and gyroscope sensors during the experiments and reasons of why we eliminated various

Table 2. Details of Users and the Data

User	Smartphone	Age range	Gender	Employment status	Data size
1	Samsung Galaxy S5	21–29	Male	Not employed	1 GB
2	-	-	-	-	666 MB
6	Samsung Galaxy Note	21–29	Male	Employed	2 GB
8	LG K7	30–39	Male	Employed	1 GB
9	LG	30–39	Female	Employed	1 GB
10	OnePlus 3	21–29	Male	Employed	6 GB
11	Samsung Core Prime	21–29	Male	Employed	1 GB
12	LG Realm	30–39	Female	Employed	301 MB
16	-	-	-	-	7 GB
19	HTC 10	30–39	Female	Employed	10 GB
20	Asus Zenfone 2	30–39	Male	Employed	6 GB
21	Samsung Galaxy S7	30–39	Male	Employed	14 GB
23	Sony Xperia Z3	60–69	Male	Retired	18 GB
24	Samsung S6 Plus Edge	21–29	Female	Employed	10 GB
26	Samsung Galaxy S6	30–39	Female	Not employed	16 GB
27	Samsung SM-A800F	21–29	Female	Employed	4 GB
28	Moto G 3rd gen	18–20	Male	Employed	845 MB
32	-	-	-	-	884 MB
36	-	-	-	-	5 GB
38	Samsung SM-A800F	30–39	Female	Employed	3 GB
39	Samsung On5	60–69	Male	Employed	1 GB

Table 3. List of sensor types and reasons for inclusion and elimination

Sensor	In-Use/Not	Reason
Accelerometer	Yes	Includes imperfection in manufacturing
Gyroscope	Yes	Includes imperfection in manufacturing
Magnetometer	No	Affected by environmental conditions
Magnetometer	No	Affected by environmental conditions
Ambient light	No	Affected by environmental conditions
Location	No	Battery consumption
Ambient temperature	No	Affected by environmental conditions
Pressure	No	Affected by environmental conditions
Humidity	No	Affected by environmental conditions

sensors, such as ambient light, location, etc. Table 3 shows the list of used and eliminated sensors during the model training along with the reasons of inclusion or elimination. Despite CrowdSignals.io provides different sensor types, we only utilize accelerometer and gyroscope sensors based on the signal imperfection during the manufacturing process, hence it makes it possible to identify devices based on the readings from these sensors.

Accelerometer and gyroscope sensors are both members of MEMS [1] (Microelectromechanical systems), which integrate mechanical and electrical components to provide sensing mechanisms on mobile devices. MEMS sensors measure linear acceleration or angular motion for one or more axis and provide an input to applications in order to detect device movements. An accelerometer senses axis orientation while gyroscope measures angular orientation for x, y and z axis, but both detect displacement of the device. Therefore, we utilize motion changes of devices to understand users' activity patterns and distinguish from each other. In addition to these sensors, we also investigate other sensor types, which exist in CrowdSignals.io dataset and we see that most of them detect ambient conditions and are affected by environmental conditions. For instance, the light sensor measures the light level and adjusts brightness of the device screen accordingly, thus the light sensor does not provide information about user behavior, unlike motion sensors. Moreover, conditions of living spaces such as humidity, pressure, etc. are mostly the same for users who share similar environments thus we focus on using motion sensors rather than others.

In addition to the environmental dependency of sensors, battery consumption and real-time performance were considered as important issues, so we also excluded location sensors from the early experiments. Moreover, observing users' location and constantly sending data to a remote system requires more Internet bandwidth which in turn impacts battery consumption. Hence, we utilize accelerometer and gyroscope sensors by considering the limited resources of the device.

3.3   Data Preprocessing

The CrowdSignals.io dataset consists of multiple files, which are categorized by sensor type and each row of these files includes a list of sensor recordings collected at the same time interval, 1 second. Firstly, we had to merge these files in the preprocessing step using the timestamp information. Secondly, since this is a streaming data, we work on windows of data to extract features, such as average, maximum, minimum, etc., from specific window sizes. We extract features for different window sizes. Reading all files for each window size caused a long preprocessing time, especially working with limited resources, such as memory size, number of CPU cores or disk capacity. Therefore, we read the data from raw files, transformed and exported as a H2O dataframe [4], thus we reduced the file sizes 10 times from the original dataset and increased the data processing speed.

3.4   Feature Extraction

Feature extraction is a critical step of machine learning process to detect informative variables, those correlated with the result, so a model can produce better results while keeping over-fitting risk as minimum. The original dataset contains streaming data of motion sensors and only have raw values from three axes of the sensors with the timestamp and label information. Accelerometer and gyroscope readings are provided in a vector which includes readings from x, y and z axes. We also calculated the magnitude value given in Eq. 1 for both sensors which makes it possible to integrate sensor information independent of the phone orientation.

$$|\vec{a}(t)| = \sqrt{a_x^2 + a_y^2 + a_z^2} \tag{1}$$

Moreover, we extracted various features, such as minimum, maximum, etc. from the readings of three axes and magnitude values. We extracted the features which are given in Table 4 for 5 (1, 2, 5, 10 and 15 seconds) different window sizes. Detecting the optimal window size for recognition is also important before deploying ML models for preventing over-consumption of mobile device resources. For instance, collecting accelerometer data once in a second may increase both Internet usage and battery usage of a mobile device due to transferring data between a mobile application and remote server constantly and sampling the sensor at a high rate. Therefore, an application should keep the balance between resource consumption of a mobile device and model performance while identifying devices.

In this paper, we did not use other features, such as zero crossing rate, skewness, kurtosis, etc., since our plan is to provide a real-time detection framework that works on the devices, considering their limited resources. Hence, we focused on simple features that are not computationally expensive. We also wanted to demonstrate the effectiveness of the detection even using simple features.

Table 4. A list of extracted features from accelerometer and gyroscope

Feature	Description	Formula
Mean	The mean of the values	$\bar{x} = \dfrac{1}{N}\displaystyle\sum_{i=1}^{N}(x_i)$
Min	The minimum of the values	$Min = \displaystyle\min_{x_1,\dots x_N}(x_i)$
Max	The maximum of the values	$Max = \displaystyle\max_{x_1,\dots x_N}(x_i)$
Standard deviation	Standard deviation of the values	$\sigma = \sqrt{\dfrac{1}{N-1}\displaystyle\sum_{i=1}^{N}(x_i-\bar{x})^2}$
Variance	Unbiased variance of the values	$\sigma^2 = \dfrac{\sum_{i=1}^{N}(x_i-\mu)^2}{N}$

3.5   Classification and Performance Metrics

In this study, we utilize three supervised classifiers (Random Forest [5], Generalized Linear Model [6] (GLM) and Gradient Boosting Machine [7] (GBM)) to identify devices based on accelerometer and gyroscope sensors. One of the reason for selecting these classifiers was that both random forest and GBM are classifiers that utilize boosting for improving detection. Random forest was used in related studies, however GBM and GLM were not evaluated in previous studies. Hence, we wanted to investigate the performance with different classifiers.

Initially, we generated various datasets for 5 different window sizes (1, 2, 5, 10 and 15 seconds) and all of them consist of labeled records from 21 users. After that, each dataset was split into 10 folds that must contain records from all users (classes), so we applied stratified sampling which assigns roundly the same percentage of records from each class similar to the entire dataset.

The entire process was the same for all classifiers but we also tested particular classifier parameters to build efficient models on training. In random forest, the number of trees, which specifies the decision tree size in the forest, can be changed. We repeated the experiments for 6 different (5, 10, 15, 20, 25, 30) tree sizes per each time window. Our aim is to detect optimal number of trees which keeps balance between model complexity and performance. In GLM, the maximum number of iterations, which specifies the nth iteration that terminates model training, can be parameterized. In the experiments, we utilized 6 different (5, 10, 15, 20, 25, 30) iterations per each time window so we were able to analyze the results with increasing number of iterations.

In the experiments, H2O.ai [8] machine learning framework is used in all phases from dataset preparation to model performance evaluation. We evaluated with the metrics those provided by H2O.ai to measure the performance of multinominal/multiclass models. We utilized accuracy and RMSE (Root Mean Squared Error) metrics which are given in Eqs. 2 and 3.

$$Accuracy = \frac{Number\ of\ correct\ predictions}{Number\ of\ total\ predictions} \tag{2}$$

Let N is the total number of rows, i represents each class, y is actual class and $\bar{y}$ is predicted class. Therefore, RMSE evaluates the size of the error per class for user identification.

$$RMSE = \sqrt{\frac{1}{N}\sum_{i=1}^{N}(y_i - \bar{y}_i)^2} \tag{3}$$

In addition to accuracy and RMSE metrics, average precision and recall are measured to evaluate overall performance of the model via a confusion matrix. Therefore, we had to calculate precision and recall per class since H2O.ai does not provide these metrics when dataset consists of more than two labels. In Eqs. 4 and 5, TP_i represents the number of true positives and FP_i is defined as the number of false positives, likewise FN_i refer to the number of false negatives for class i.

$$precision_i = \frac{(TP_i)}{(TP_i + (FP_i)} \tag{4}$$

$$recall_i = \frac{(TP_i)}{(TP_i + (FN_i)} \tag{5}$$

As the last step, we calculated average precision and recall in order to evaluate the model performance as follows. Let m is the total number of classes:

$$Avg\ Precision = \frac{\sum_{i=1}^{m} precision_i}{m} \tag{6}$$

$$Avg\ Recall = \frac{\sum_{i=1}^{m} recall_i}{m} \tag{7}$$

4   Performance Evaluation

In this section, we present the results of the experiments. As discussed in Sect. 3, we have different parameters that are considered in the experiments and full list of parameters is given in Table 5. Considering different sensors (2 sensors and combination), five different window sizes or number of iterations, 90 models were built for all classifiers.

In the following subsections, we present the results per sensor and their combination, considering the impact of different parameters. We start with the results obtained with the accelerometer sensor, and next we provide the results of the gyroscope sensor, followed by their combination. Results of each experiment are presented in terms of accuracy, RMSE, precision and recall.

Table 5. Experiment setup parameters

Parameter	Values	Extras
Classifiers	GBM, GLM, Random Forest	Random Forest, GBM: 5, 10, 15, 20, 25, 30 trees GLM: 5, 10, 15, 20, 25, 30 iterations
Features	Mean, min, max, standard deviation, variance	Calculated from X, Y, Z axes and magnitude for each sensor
Window Sizes	1, 2, 5, 10 and 15 seconds	
Sensors	Accelerometer, gyroscope, combination	

4.1   Accelerometer

In this section, we aim to analyze the effects of window size and classifier on the performance of the model which was created by the data from only accelerometer sensor. In Fig. 1, we compare the performance results of Random Forest, GBM and GLM classifiers for 5 different window sizes which are presented in Table 5. The x-axis shows the window size whereas y-axis displays the accuracy values, varying between 0 and 1. When we compare the accuracy results, GBM obtains the highest accuracy as 0.91 on 1 and 2 second windows with a small difference. The results show that Random Forest also works with accelerometer sensor efficiently as well as GBM and classifies 89% of the devices correctly. When the results of GLM are compared to others, we obtain the lowest accuracy as 0.31 on average considering every window size. The reason is that GLM is more convenient to handle simple linear problems which can be analytically solved rather than complex ones. Furthermore, we inferred that increasing the window size causes a slight decrease after 2 seconds for Random Forest and GBM classifiers. As a result, GBM is observed to be the best performing classifier to identify devices when compared to others.

4.2   Gyroscope

In this section, we evaluate the impact of both the classifier and the window size on model performance as in Sect. 4.1 but here we use the gyroscope sensor. In Fig. 2, y-axis shows the accuracy as a performance metric and x-axis displays the window size. The highest accuracy is 91%, obtained by the GBM classifier on 1 second window and Random Forest achieved the second best accuracy 83% in almost every window size. However, GLM exhibits the lowest accuracy in every window size, maximum 31%, same as the results in Sect. 4.1. The performance of GBM models were affected positively by increasing window size starting with 2 seconds whereas Random Forest stays the same. However, 2 second windows led to significant increase on the performance of GLM classifier and enhanced accuracy from 20% to 30%. In conclusion, GBM provides the best accuracy, 91% same as the accelerometer on 1 second window and GLM is inadequate to classify devices for any window size.

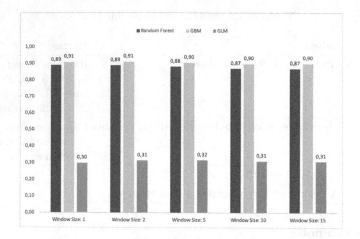

Fig. 1. Accuracy with accelerometer with 3 classifiers and 5 different window sizes

4.3   Accelerometer and Gyroscope

In this section, our aim is to explore the impact of combining accelerometer and gyroscope features that are mentioned in Table 4 on the model performance. Figure 3 shows that all the classifiers exhibit better results while using both sensors together. Again, GBM is the most efficient classifier with the highest accuracy as 98% with 15 second window size. However, increasing window size causes a decrease in the number of instances for some users since some users have less data. For this reason, 5 second window was utilized with the aim of comparing the performance of sensors reliably in Fig. 4. When we consider different window sizes from 1 to 10 in Fig. 3, GBM provides the highest accuracy,

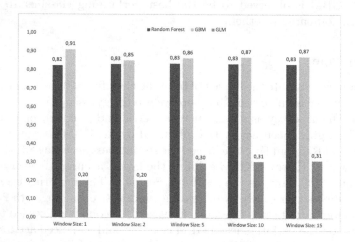

Fig. 2. Accuracy with gyroscope with 3 classifiers and 5 different window sizes

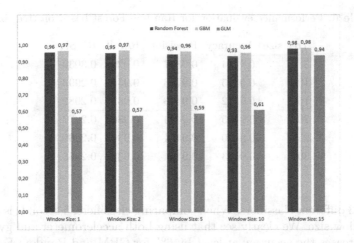

Fig. 3. Accuracy with accelerometer and gyroscope with 3 classifiers and 5 different window sizes

97% on 1 second window. Besides, the performance of Random Forest is slightly smaller than GBM: 96% is the maximum accuracy of all the Random Forest models. Furthermore, the worst performance is obtained by GLM once again but better than the previous GLM models created by using accelerometer or gyroscope.

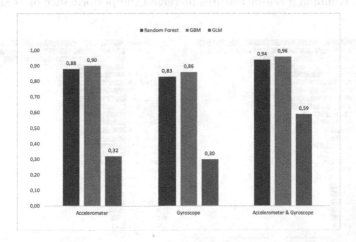

Fig. 4. Accuracy comparison for 3 different classifiers according to sensor type

Finally, we compare the performance of classifiers by using accelerometer, gyroscope and their combination in Fig. 4. The y-axis represents accuracy as in the previous graphs and x-axis shows which sensor(s) is used while training

Table 6. Performance evaluation for Random Forest based on tree size

Tree size	Accuracy	Precision	Recall	RMSE
5	0.9334	0.9337	0.9258	0.3030
10	0.9395	0.9403	0.9325	0.2993
15	0.9412	0.9428	0.9344	0.2982
20	0.9425	0.9444	0.9360	0.2977
25	0.9430	0.9451	0.9365	0.2965
30	0.9433	0.9454	0.9368	0.2955

models by 3 different classifiers. When comparing classifiers, 5 seconds is selected as the window size. We clearly see that using both accelerometer and gyroscope sensors increases the accuracy at least by 6% for GBM and Random Forest.

Additionally, we also utilized several performance metrics in Table 6 such as precision, recall, etc. mentioned in Sect. 3.5. Table 6 shows the performance evaluation of Random Forest model for 6 different tree size. The results demonstrate that increasing tree size enhances accuracy, precision and recall metrics and improves all of them 1% in average. In addition to these improvements, RMSE decreases at every step so model classifies 0.75% more devices correctly by increasing tree size from 5 to 30.

Lastly, we analyzed the importance of features in Figs. 5, 6 and 7 for all classifiers which are using accelerometer and gyroscope sensors together. As a conclusion, combining sensor features increases the performance of classifiers.

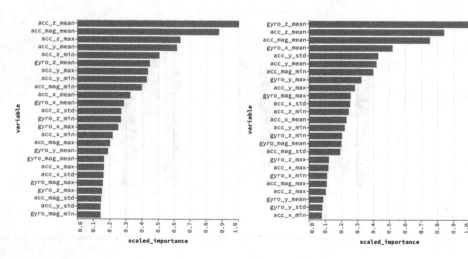

Fig. 5. Random Forest feature importance for 5 second window

Fig. 6. GBM feature importance for 5 second window

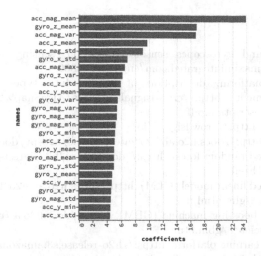

Fig. 7. GLM feature importance for 5 second window

5 Conclusion and Future Work

In this paper, we investigated user/device identification using sensor fingerprinting particularly on mobile devices. We particularly investigated the performance
of identification using two motion sensors, accelerometer and gyroscope which
are commonly available on smart phones using a set of simple features. We utilized a large dataset, named as CrowdSignals.io, which is larger than 25 GB,
and consists of sensor data from 21 distinct devices. We also investigated the
performance with three different classification methods, namely random forest,
gradient boosting machine, and generalized linear model using 40 features, also
considering the effect of different time window sizes. Experiment results show
that random forest and GBM classifier exhibit similar performances, while GLM
performs much worse than the others. When only accelerometer is used, GBM
achieves 91% accuracy on 1 and 2 second time windows, while random forest
performs with 89% accuracy in these cases, and GLM with 31%. Similar results
were achieved with different window sizes. When gyroscope is used, slightly lower
results were achieved, ranging between 82 to 91% accuracy for random forest and
GBM classifiers. However, GBM with 1 second window size exhibited a similar
performance as the gyroscope case: 91% accuracy. When both sensors are considered, accuracy increases to 97% for GBM and 96% for random forest. As a future
work, we plan to investigate the performance with a smaller number of features
and considering different sensors. We also plan to apply deep learning methods
to compare the findings with the traditional machine learning algorithms.

Acknowledgements. This work is supported by the Galatasaray University Research
Fund under grant number 17.401.004 and by Tubitak under grant number 5170078.

References

1. Accelerometer and gyroscopes sensors: operation, sensing, and applications. https://pdfserv.maximintegrated.com/en/an/AN5830.pdf
2. Crowdsignals platform, description of sensor data types collected, user's reference document. http://crowdsignals.io/docs/AlgoSnap%20Sensor%20Data%20Types%20Description.pdf
3. Crowdsignals.io. http://crowdsignals.io/
4. Data in h2o.ai. http://docs.h2o.ai/h2o/latest-stable/h2o-py/docs/data.html
5. H2o.ai distributed random forest. http://docs.h2o.ai/h2o/latest-stable/h2o-docs/data-science/drf.html
6. H2o.ai generalized linear model (GLM). http://docs.h2o.ai/h2o/latest-stable/h2o-docs/data-science/glm.html
7. H2o.ai gradient boosting machine (GBM). http://docs.h2o.ai/h2o/latest-stable/h2o-docs/data-science/gbm.html
8. H2o.ai machine learning platform. https://h2o-release.s3.amazonaws.com/h2o/rel-slater/9/docs-website/h2o-py/docs/index.html
9. Bojinov, H., Michalevsky, Y., Nakibly, G., Boneh, D.: Mobile device identification via sensor fingerprinting. CoRR abs/1408.1416 (2014)
10. Das, A., Borisov, N., Caesar, M.: Tracking mobile web users through motion sensors: attacks and defenses. In: 23rd Annual Network and Distributed System Security Symposium, NDSS 2016, San Diego, California, USA, 21–24 February 2016. http://wp.internetsociety.org/ndss/wp-content/uploads/sites/25/2017/09/tracking-mobile-web-users-through-motion-sensors-attacks-defenses.pdf
11. Dey, S., Roy, N., Xu, W., Choudhury, R.R., Nelakuditi, S.: AccelPrint: imperfections of accelerometers make smartphones trackable. In: 21st Annual Network and Distributed System Security Symposium, NDSS 2014, San Diego, California, USA, 23–26 February 2014. https://www.ndss-symposium.org/ndss2014/accelprint-imperfections-accelerometers-make-smartphones-trackable
12. Ehatisham-ul Haq, M., et al.: Authentication of smartphone users based on activity recognition and mobile sensing. Sensors **17**(9) (2017). https://doi.org/10.3390/s17092043. https://www.mdpi.com/1424-8220/17/9/2043
13. Lee, W., Lee, R.B.: Sensor-based implicit authentication of smartphone users. In: 2017 47th Annual IEEE/IFIP International Conference on Dependable Systems and Networks (DSN), pp. 309–320, June 2017. https://doi.org/10.1109/DSN.2017.21
14. Mayrhofer, R., Gellersen, H.: Shake well before use: authentication based on accelerometer data. In: LaMarca, A., Langheinrich, M., Truong, K.N. (eds.) Pervasive 2007. LNCS, vol. 4480, pp. 144–161. Springer, Heidelberg (2007). https://doi.org/10.1007/978-3-540-72037-9_9
15. Welbourne, E., Munguia Tapia, E.: CrowdSignals: a call to crowdfund the community's largest mobile dataset, pp. 873–877, September 2014. https://doi.org/10.1145/2638728.2641309

A SAT-Based Formal Approach for Verifying Business Process Configuration

Abderrahim Ait Wakrime[1(✉)], Souha Boubaker[2], Slim Kallel[3], and Walid Gaaloul[2]

[1] Computer Science Department, Mohammed V University, Rabat, Morocco
abderrahim.aitwakrime@gmail.com
[2] Computer Science Department, SAMOVAR, Telecom SudParis, CNRS,
Paris-Saclay University, Evry, France
[3] ReDCAD, University of Sfax, Sfax, Tunisia

Abstract. Nowadays, some organizations are using and deploying similar business processes to achieve their business objectives. Typically, these processes often exhibit specific differences in terms of structure and context depending on the organizations needs. In this context, configurable process models are used to represent variants in a generic manner. Hence, the behavior of similar variants is grouped in a single model holding configurable elements. Such elements are then customized and configured depending on specific needs. Nevertheless, the decision to configure an element may be incorrect leading to critical behavioral errors. In the present work, we propose a formal model based on propositional satisfiability formula allowing to find all possible correct elements configuration. This approach allows to provide the designers with correct configuration decisions. In order to show the feasibility of the proposed approach, an experimentation was conducted using a case study.

Keywords: Configurable Business Process · Formal methods ·
Propositional satisfiability

1 Introduction

Configurable business process models are increasingly adopted by companies due to their capability of grouping the common and variable parts of similar processes. They allow to integrate multiple process variants of a same business process in a given domain. The differences between these variants are captured through some variation points allowing for multiple design options. These points are referred to as configurable elements that can be configured and adjusted according to each specific organization requirement by selecting the desired options and deselecting the undesired ones. The resulting processes are called variants.

© Springer Nature Switzerland AG 2019
M. Younas et al. (Eds.): Innovate-Data 2019, CCIS 1054, pp. 47–62, 2019.
https://doi.org/10.1007/978-3-030-27355-2_4

Since the configurable elements may have large number of configuration options with complex inter-dependencies between them, the decision of manually applying correct ones is tedious and error-prone task. We mean by correct options, those that allow to generate correct process variants without structural and execution errors.

Different approaches have been proposed to deal with process configuration. Some of them aimed to facilitate the configuration of process models [1–4]. While other approaches proposed to guide the configuration and to support domain-based constraints [5–8]. Others attempted to ensure the process configuration correctness [9, 10]. However, these approaches mainly suffer from the state space explosion problem.

In this paper, we propose to address this issue and to further improve the efficiency of business process configuration by using the satisfiability problem (SAT). SAT studies Boolean functions, and s especially concerned with the set of truth assignments (assignments of 0 or 1 to each of the variables) making the function true. Several research works have used SAT in different application domains to effectively solve a set of problems and obtain good performance, like for example: theorem proving, planning, non-monotonic reasoning, electronic design automation or knowledge-base verification and validation [11–15]. Thanks to the advance in SAT resolution, SAT solvers are recently becoming the tool for tackling more and more practical problems.

Our main goal is to apply the satisfiability problem in the analysis and the verification of configurable process models. For this aim, we propose a SAT-based formal approach that allows to generate all correct configurations of a configurable process model. These correct options help and assist the process designer to easily identify correct process variants.

Concretely, in this paper we provide translation rules of a configurable business process into SAT model in order to analyze and verify how a process model's behavior can be restricted. We define a formal model that allows to derive the different configurations leading to correct variants.

The rest of the paper is structured as follows. In Sect. 2, we motivate our approach using an example of configurable process model and we give some preliminaries about configurable business process and propositional satisfiability. Section 3 illustrates our formalization of process configuration. The approach validation is depicted in Sect. 4. We present the related work in Sect. 5. Finally, we conclude and provide insights for future work.

2   Motivating Example and Background

In this section, we present first a motivating example that illustrate the usefulness of the problem. Second, we review the basic notions of propositional satisfiability.

2.1 Motivating Example: Configurable Business Process

In Fig. 1, we present a simplified example of a configurable process model designed by a process provider for a hotel booking agency. The process is modeled using the Configurable Business Process Model and Notation (C-BPMN) [3,16], a configurable extension to BPMN[1].

We consider four main control flow elements: activity (represented with a rectangle), edge (represented with arrows), event (represented with a circle) and connector (represented with a diamond). Three main connectors are used to model the splits (e.g. s_1) and the joins (e.g. j_1) : OR ($\bigcirc$), exclusive OR ($\times$) and AND ($+$).

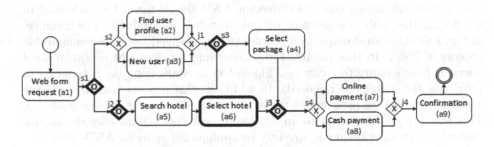

Fig. 1. Hotel booking configurable process example.

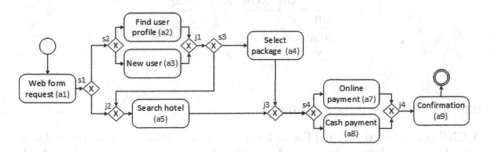

Fig. 2. A process variant derived from the configurable process in Fig. 1

Since a configurable process can not be executed, all configurable elements should be configured and customized in order to obtain a variant that can be instantiated. A C-BPMN process includes two types of configurable elements: *activities* and *connectors*. This example presents 5 configurable elements (4 connectors and one activity) which are highlighted with a thicker border. The non-configurable ones represent the commonalities in the configurable model. This travel agency has a number of branches in different countries. Depending on

---

[1] BPMN 2.0 specification http://www.omg.org/spec/BPMN/2.0/.

specific needs of a country, each branch performs a different variant of this process model in terms of structure and behavior. For instance, activities a_1 and a_8 are non-configurable, which means that they should be included in every configured variant. Whereas, the activity a_6 and the connector s_1 may vary from one process to another, as they are configurable.

A connector may be configurable to restrict its behavior. It can be configured by (i) changing its type (e.g. from OR to AND), or/and (ii) restricting its incoming or outgoing branches. By configuration, a connector may change its type according to a set of configuration constraints [1] (see Table 1). Each row corresponds to the initial type that can be mapped and configured to one or more types in the columns. For example, a configurable connector having the OR type can be configured to any type while an AND type remains unchangeable. It is worth noting that the connector AND should never be configured to a sequence (i.e., only one input or output branch is maintained). For instance, in Fig. 2 we show an example of a process variant derived from the configurable process in Fig. 1. In this variant, the process analyst does not need the hotel selection functionality (activity a_6). This refers to configuring the activity a_6 to *OFF* (i.e. the activity is removed). In addition, this process tenant needs the execution of only one path among the outputs of s_1 (i.e. configurable connector). This refers to configuring s_1 to an XOR-split. Another tenant may choose, for example, to execute them concurrently by configuring s_1 to an AND-split.

Table 1. Constraints for connectors configuration [1].

	OR	XOR	AND	SEQ
OR	✓	✓	✓	✓
XOR		✓		✓
AND			✓	

2.2   Propositional Satisfiability

A CNF (Conjunctive Normal Form) formula Σ is a conjunction (interpreted as a set) of clauses, where a clause is a disjunction (interpreted as a set) of literals. A literal is a positive (x) or negative ($\neg x$) Boolean variable. The two literals x and $\neg x$ are called complementary. An unit clause is a clause with only one literal (called unit literal). An empty clause, is interpreted as false, while an empty CNF formula, is interpreted as true. A set of literals is complete if it contains one literal for each variable occurring in Σ and fundamental if it does not contain complementary literals. An interpretation $\mathcal{I}$ of a Boolean formula Σ associates a value $\mathcal{I}(x)$ to some of the variables x appearing in Σ. An interpretation can be represented by a fundamental set of literals, in the obvious way. A model of a formula Σ is an interpretation $\mathcal{I}$ that satisfies the formula, *i.e.* that satisfies all clauses of the formula. SAT is the problem of deciding whether a given CNF formula Σ admits a model or not. $\models_*$ denotes the logical consequence modulo unit propagation. Any propositional formula can be translated into an equi-satisfiable formula in CNF using Tseitin's linear encoding [17].

Let us now briefly describe the basic components of CDCL (Conflict-Driven Clause Learning)-based SAT solvers [18,19]. To be exhaustive, these solvers incorporate unit propagation (enhanced by efficient and lazy data structures), variable activity-based heuristic, literal polarity phase, clause learning, restarts and a learnt clauses database reduction policy. Typically, a SAT solver can be assimilated to a sequence of decision and unit propagation literals. Each literal chosen as a decision variable is affected to a new decision. If all literals are assigned, then $\mathcal{I}$ is a model of the formula and the formula is answered to be satisfiable. If a conflict is reached by unit propagation, a new clause is derived by conflict analysis [20] considered as a logical consequence of the initial problem. If an empty clause is derived, then the formula is answered to be unsatisfiable.

3   SAT-Based Business Process Configuration

In this section, we introduce our SAT-based approach for business process configuration using a formalism for modeling the behavior of all possible variants of a configurable process model.

3.1   Our Approach in a Nutshell

Our approach consists of three steps (see Fig. 3). In the first step, a configurable business process is transformed into a SAT formalism using propositional logic. The second step corresponds to finding all correct configurations. Typically, this consists in generating the set of all combinations of elements configurations

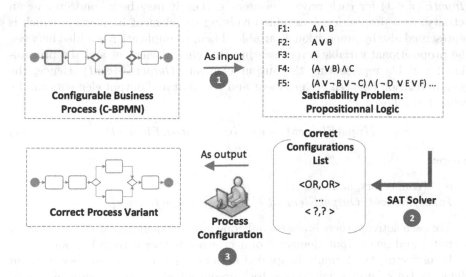

Fig. 3. Our approach overview.

(i.e., connectors and activities configurations) leading to correct process variants. This means that the obtained process models by applying each obtained combination of elements configurations should satisfy the formula obtained in the first step. These combinations are generated using a SAT solver.

Once the correct configurations are obtained, the process analysts will be able to correctly choose and configure their process variant without additional checking.

3.2 Formalizing Business Process Configuration Using SAT

In the following, we define our formal approach for representing the control flow elements of a Business Process (BP) as well as a Configurable Business Process (CBP) using SAT.

In our SAT-based formal model, each element of a Business Process, namely activities and connectors, is transformed to a propositional formula using propositional variables and logical connectors (like: $\neg, \vee, \wedge, \rightarrow$). The selection and generation of the BPs variants during the configuration are related to a conditional statement or a conditional expression and they are represented as a simple implication $(p \rightarrow q)$ in classical logic in a SAT. This implication is read as (if p then q). It merely means (if p is true, then q is also true). The statement $(p \rightarrow q)$ is false only when p is true and q is false.

The control flow perspective describes activities and their execution ordering through different constructors, which permit its execution [21]. This control flow is ensured by edges that indicates the execution direction (to the right). Each node or element (that can be an activity or a connector) have one or more inputs and one or more outputs.

To formally represent the control flow, we firstly represent the input element (*InputElement*) for each process element **e** (i.e., it may be a connector or an activity) as a propositional variable involving the element in question which is represented also by propositional variable. Then, an implication is added between the propositional variable **e** representing this current element and the propositional variable representing the output element (*OutputElement*). Hence, the general relation between each element and its input and output elements can be defined as follows:

$$(InputElement \rightarrow \mathbf{e}) \wedge (\mathbf{e} \rightarrow OutputElement) \tag{1}$$

where:

- $\mathbf{e} \in \{\text{connectors, activities}\}$;
- $InputElement, OutputElement \in \{connectors, activities\}$.

For each activity, there is always only one input element (connector or another activity) and one output element (connector or another activity). However, for each connector, the formula is applied as much times as the number of input elements (in case of a join connector) or output elements (in case of a split connector). The implication (1) can be easily translated into the following clause in order to obtain CNF formula:

$$\psi \; = \; (\neg InputElement \vee e) \wedge (\neg e \vee OutputElement) \tag{2}$$

In the following, we use the formula ψ in order to translate a configurable process to classical logic in a SAT, then to CNF. However, prior to that, we define in Table 2 the formalization of every configurable connector type.

Table 2 depicts the mapping from a set of BPMN activities and connectors to SAT formulas. The first column contains the BPMN elements, the second one represents the SAT configurable connectors formalization and the third one contains SAT non-configurable connectors formalization. These mappings, as shown in Table 2, are straightforward. A connector is mapped onto a disjunction between two implications with their input activities and their output activities. The first implication represents the relation between the connector and their input activities, but, the second one represents the relation between the connector and their output activities. Each connector and each activity is formalised using a propositional variable. The transformation is based on type of each connector which could be OR, XOR or AND with their two variants namely join or split.

In order to obtain a correct configuration, we start by defining a variable $Activity_{root}$ that represents the initial activity and we conjunctively add the formula ψ as follows: $\psi \wedge Activity_{root}$. We also define the variable $Activity_{target}$ that represents the final activity in the process. Then, a correct process variant is a configuration that considers the initial activity as a starting point and applying unit propagation, that is a rule of correct inference, to reach the final activity. In this work, we artificially add initial and final activities to respectively represent the unique initial point of the process model and the unique final one. The Definition 1 explains how to extract all correct configurations.

Definition 1 (Correct Configuration). *Let cbp a C-BPMN and let ψ be a SAT formula that represents a transformation of cbp to a CNF formula. A configuration conf of cbp is a process where each element $e \in conf$ allows the following formula:*

$$\psi \wedge Activity_{root} \models_* Activity_{target}$$
$$which \; means \; that: \psi \wedge Activity_{root} \wedge \neg Activity_{target} \models_* \bot.$$

Hence, the Definition 1 implies that a configurable business process have one or more correct configurations *iff* the formula $\psi \wedge Activity_{root} \models_* Activity_{target}$ admits one or more process models. This means that a correct process is possible *iff* from $\psi \wedge Activity_{root}$, we can deduce by *unit propagation* (also called *Boolean constraint propagation*) the $Activity_{target}$. Furthermore, all possible correct configurations can be extracted using unit propagation. This later is an automated theorem proving procedure used to simplify a set of clauses. Also, it is one of the key processes and the most used one in SAT resolution algorithms. Its working principle is the following: until that formula contains a unitary clause, affect its literal in true. That is specified by the following definition.

Definition 2 (Correct Workflow). *Let cbp a C-BPMN. A unit propagation allows to obtain a correct process starting from the initial activity $Activity_{root}$ to the final activity $Activity_{target}$.*

The proposed relations between each connector and its input and output elements are resumed in Table 2. The connector's type is represented in the first column. For each type, the SAT formulas representing the control flow relations including this type, either configurable or not, are defined in the second and the third columns. We obtain two rules for each type.

Note 1. In the table, Cn refers to the configurable connector in the second column and to the regular connector (i.e., non-configurable) in the third column. The input element is depicted by in_k and the output element by out_k.

For instance, to formalize the split connector having OR type we define the formula: $((\bigvee_{in_k \in in_{Or_{jx}}} in_k) \to Cn) \wedge (Cn \to out)$. This implication means that, during the execution of this connector Cn either one or several input elements in_k are executed, and its execution leads to the firing of the output element out. This rule is relevant for both configurable and non-configurable (regular) connectors.

Algorithm 1 represents our algorithm that formalizes each configurable business process. This Algorithm is based on the formulas of Table 2. First, the formula Φ and ψ are declared and are initialed (lines 1 and 2). Second, In the first *foreach* loop and for each configurable connector, we apply $R_{C\_SAT}()$ to formalize the connector in question (lines 3 and 4). It is the same for others *foreach* loops, *i.e.* non-configurable connectors (lines 5 and 6) and configurable activities (lines 7, 8 and 9) respectively with $R_{NC\_SAT}()$ and $R_{CA\_SAT}()$. The formula ψ is defined by all *foreach* remaining loops. Finally, the formula Φ is defined by the conjunction between ψ, $Activity_{root}$ and $\neg Activity_{target}$, then it is returned (line 11).

Note 2. $Pre(Act)$ represents the input activity of activity Act. Similarly, $Post(Act)$ represents the output activity of activity Act.

Let us consider the set of activities and connectors as depicted in configurable business process in Fig. 1. The rules in Table 2 are applied to this configurable business process with its non-configurable connectors like s_2, j_1, s_4, j_4, its configurable connectors like s_1, s_3, j_2, j_3, its non-configurable activities: $a_1, a_2, a_3, a_4, a_5, a_7, a_8, a_9$ and its configurable activity like a_6. Furthermore, if the Algorithm 1 is applied, the following formulas ψ is obtained:

$$\psi = (a_1 \to s_1) \wedge s_1 \to (s_2 \vee j_2) \wedge$$
$$(s_1 \to s_2) \wedge (s_2 \to (a_2 \wedge \neg a_3) \vee s_2 \to (a_3 \wedge \neg a_2)) \wedge$$
$$((s_1 \vee s_3) \to j_2) \wedge (j_2 \to a_5) \wedge$$
$$((a_2 \wedge \neg a_3) \to j_1 \vee (a_3 \wedge \neg a_2) \to j_1) \wedge (j_1 \to s_3) \wedge$$
$$(j_1 \to s_3) \wedge (s_3 \to (a_4 \vee j_2)) \wedge$$
$$((a_4 \vee a_6) \to j_3) \wedge (j_3 \to s_4) \wedge$$

Table 2. Different configurable/non-configurable connectors of Business processes and the corresponding SAT formulas.

BPMN elements		SAT configurable connectors	SAT non-configurable connectors
OR-join		$((\bigvee_{in_k \in in_{Or_{jx}}} in_k) \to Cn) \wedge (Cn \to out)$	$((\bigvee_{in_k \in in_{Or_{jx}}} in_k) \to Cn) \wedge (Cn \to out)$
OR-split		$(in \to Cn) \wedge (Cn \to (\bigvee_{out_k \in out_{Or_{sx}}} out_k))$	$(in \to Cn) \wedge (Cn \to (\bigvee_{out_k \in out_{Or_{sx}}} out_k))$
AND-join		$((\bigvee_{k \in \{1...|in_{And_{jx}}|\}} \bigwedge_{i \in \{1...k\}} in_i) \to Cn) \wedge (Cn \to out)$	$((\bigwedge_{in_k \in in_{And_{jx}}} in_k) \to Cn) \wedge (Cn \to out)$
AND-split		$(in \to Cn) \wedge (Cn \to (\bigvee_{j \in \{1...|out_{And_{sx}}|\}} \bigwedge_{i \in \{1...j\}} out_i))$	$(in \to Cn) \wedge (Cn \to (\bigwedge_{out_k \in out_{And_{sx}}} out_k))$
XOR-join		$((\bigvee_{in_k \in in_{Xor_{jx}}} in_k) \to Cn) \wedge (Cn \to out)$	$\bigvee_{in_r \in in_{Xor_{jx}} \wedge in_k \in in_{Xor_{jx}} \setminus \{in_r\}} (in_r \wedge \neg in_k) \to Cn) \wedge (Cn \to out)$
XOR-split		$(in \to Cn) \wedge (Cn \to (\bigvee_{out_k \in out_{Or_{sx}}} out_k))$	$(in \to Cn) \wedge (Cn \to \bigvee_{out_r \in out_{Xor_{jx}} \wedge out_k \in out_{Xor_{jx}} \setminus \{out_r\}} (out_r \wedge \neg out_k))$

$$(j_3 \rightarrow s_4) \wedge (s_4 \rightarrow (a_7 \wedge \neg a_8) \vee s_4 \rightarrow (a_8 \wedge \neg a_7)) \wedge$$
$$((a_7 \wedge \neg a_8) \rightarrow j_4 \vee (a_8 \wedge \neg a_7) \rightarrow j_4) \wedge (j_4 \rightarrow a_8)) \wedge$$
$$(a_5 \rightarrow (a_6 \wedge \neg j_3) \vee a_5 \rightarrow (j_3 \wedge \neg a_6)) \wedge (j_2 \rightarrow a_5) \wedge$$
$$(a_1 \rightarrow s_1) \wedge$$
$$(s_2 \rightarrow a_2) \wedge (a_2 \rightarrow j_1) \wedge$$
$$(s_2 \rightarrow a_3) \wedge (a_3 \rightarrow j_1) \wedge$$
$$(s_3 \rightarrow a_4) \wedge (a_4 \rightarrow j_3) \wedge$$
$$(j_2 \rightarrow a_5) \wedge (a_5 \rightarrow a_6) \wedge$$
$$(s_4 \rightarrow a_7) \wedge (a_7 \rightarrow j_4) \wedge$$
$$(s_4 \rightarrow a_8) \wedge (a_8 \rightarrow j_4) \wedge$$
$$(j_4 \rightarrow a_9)$$

Therefore, the formula Φ is: $\Phi = \psi \wedge a_1 \wedge \neg a_9$.

4   Evaluation: SAT Problems Induced by Business Process

In this section, we are interested in experiment to verify the efficiency of the approach presented in this paper. To do so, we implemented Algorithm 1 described above. The approach has be implemented and tested on a real configurable business process illustrated by Fig. 1. Our approach has as input this configurable business process. It generates afterward the SAT encoding represented by formula Φ and is represented under a file having for extension *.dimacs*. Each line in this file is a list of variables separated by spaces and ended with 0. This list represents a clause which is a disjunction of literals and a literal is either a positive variable x or its negation $\neg x$. In the file *.dimacs*, a variable is represented by an integer between 1 and n and its negation $\neg$ is represented by the sign $-$. Indeed, the lines of a file *.dimacs* represent the conjunction of the clauses of the problem. The first line of a file *.dimacs* is written as following:

Algorithm 1. Configurable Business Processes to SAT

    Input: Configurable business process
    Output: Formula Φ that represent a SAT formalization
1   Φ is a SAT formula;
2   ψ is a SAT formula;
3   **foreach** $Conn \in \{configurable\ connectors\}$ **do**
4    $\lfloor$ $\psi \leftarrow R_{C\_SAT}(Conn)$;
5   **foreach** $Conn \in \{non\text{-}configurable\ connectors\}$ **do**
6    $\lfloor$ $\psi \leftarrow R_{NC\_SAT}(Conn)$;
7   **foreach** $Act \in \{configurable\ activities\}$ **do**
8    $|$ $Act \leftarrow Pre(Act)$;
9    $\lfloor$ $\psi \leftarrow R_{CA\_SAT}(Act)$;
10 $\Phi \leftarrow \psi \wedge Activity_{root} \wedge \neg Activity_{target}$;
11 **return** Φ

$$p \quad cnf \quad n \quad c$$

This line indicates that the instance is in CNF format with n is the number of variables that represent the activities and connectors and c is the number of clauses of the problem. The file *.dimacs* generated by the Algorithm 1 is then introduced into the SAT solver MiniSat[2]. This SAT solver is based on Boolean satisfiability problem that is the decision problem whether a propositional formula is evaluated to *true* for any assignment of its variables. In this case, the formula is satisfiable (SAT), otherwise the formula is unsatisfiable (UNSAT). The Fig. 4 represents an extract of *.dimacs* file concerning Fig. 1. In Fig. 4, each line represents a clause that is a sequence of distinct non-null numbers between $-n$ and n ending with 0 on the same line. The opposite literals x and $-x$ do not

Table 3. All correct configurations

s_1	s_3	j_2	j_3	✓ : including activity a_6 / a_6	s_1	s_3	j_2	j_3	✗ : without activity a_6 / a_6
OR	OR	OR	OR	✓	OR	OR	OR	OR	✗
XOR	OR	OR	OR	✓	XOR	OR	OR	OR	✗
AND	OR	OR	OR	✓	AND	OR	OR	OR	✗
OR	XOR	OR	OR	✓	OR	XOR	OR	OR	✗
XOR	XOR	OR	OR	✓	XOR	XOR	OR	OR	✗
AND	XOR	OR	OR	✓	AND	XOR	OR	OR	✗
OR	AND	OR	OR	✓	OR	AND	OR	OR	✗
XOR	AND	OR	OR	✓	XOR	AND	OR	OR	✗
AND	AND	OR	OR	✓	AND	AND	OR	OR	✗
OR	OR	XOR	OR	✓	OR	OR	XOR	OR	✗
XOR	OR	XOR	OR	✓	XOR	OR	XOR	OR	✗
OR	XOR	XOR	OR	✓	OR	XOR	XOR	OR	✗
XOR	XOR	XOR	OR	✓	XOR	XOR	XOR	OR	✗
OR	AND	AND	OR	✓	OR	AND	AND	OR	✗
AND	AND	AND	OR	✓	AND	AND	AND	OR	✗
OR	OR	OR	XOR	✓	OR	OR	OR	XOR	✗
XOR	OR	OR	XOR	✓	XOR	OR	OR	XOR	✗
OR	XOR	OR	XOR	✓	OR	XOR	OR	XOR	✗
XOR	XOR	OR	XOR	✓	XOR	XOR	OR	XOR	✗
OR	OR	XOR	XOR	✓	OR	OR	XOR	XOR	✗
XOR	OR	XOR	XOR	✓	XOR	OR	XOR	XOR	✗
OR	XOR	XOR	XOR	✓	OR	XOR	XOR	XOR	✗
XOR	XOR	XOR	XOR	✓	XOR	XOR	XOR	XOR	✗
AND	AND	OR	AND	✓	AND	AND	OR	AND	✗
AND	AND	AND	AND	✓	AND	AND	AND	AND	✗

---

[2] http://minisat.se/.

$$
\begin{array}{cccc}
p & cnf & 17 & 43 \\
-1 & 2 & 0 & \\
-2 & 3 & 4 & 0 \\
-2 & 3 & 0 & \\
-3 & 5 & 6 & 0 \\
\end{array}
$$
....

Fig. 4. An extract of *.dimacs* file with 17 variables and 43 clauses.

belong to the same line. In addition, positive number denotes the corresponding variable and the negative one denotes the negation of the corresponding variable.

The SAT solver will check the existence of the correct configurations and it will generate them if they exist. Else, the SAT solver returns UNSAT *i.e.* the configurable business process does not contain any correct configuration.

In order to evaluate our approach illustrated by Fig. 3, we conducted tests on a 64-bit PC, Ubuntu 16.04 operating system, an Intel Core i5, 2.3 GHz Processor with 4 cores and 8 GB RAM. Table 3 represents the different correct configurations. These correct configurations are represented and classified using the configurable activity a_6. In the left of this table, all correct configurations including the activity a_6 are presented. It is the same for all correct configurations without the activity a_6 which are presented in the right of this table. As explained, our approach always aims directly the correct configurations when available.

5   Related Work

Several approaches have been proposed to model variability and to provide a correctness verification of the configurable process models [1,3,4,6,7,22–27]. In this context and in [22], Gottschalk et al. propose an approach for extending YAWL language, as a common workflow modelling language with opportunities for predefining alternative model versions within a single workflow model. They propose to allow the configuration of workflow models to a relevant variant in a controlled way.

Schunselaar et al. propose in [23] an approach to capture the control-flow of a process defined by a CoSeNet, which is a configurable and tree-like representation of the process model. They focus also on presenting how to merge two CoSeNets into another CoSeNet such that the merge is reversible.

Configurable Event-Driven Process Chains (C-EPCs) [1] is an extended reference modelling language which allows capturing the core configuration patterns. The authors define the formalization of C-EPCs as well as examples for typical configurations. In addition, they propose the identification of a comprehensive list of configuration patterns and they test the quality of these extensions in some experiments.

van der Aalst et al. [25] propose an approach inspired by the "operating guidelines" used for partner synthesis for verifying that configuration do not lead to behavioral issues like deadlocks and livelocks. They represent the configuration process as an external service, and compute a characterization of all such services which meet particular requirements via the notion of configuration guideline. [26] proposes an approach for process family architecture modeling and implementation. The authors propose a set of variability mechanisms for BPMN and outlined their implementation using HyperSenses program generators. In the approach [3], the modeling of a reference process model which represent a base process model is discussed. The necessary adjustments of this process are treated to configure this base process model to different process variants. This is done by introducing the Provop framework.

Other authors focused also on the issue of process models configuration. For example, in [6] the different variants of configurable process models are derived based on domain constraints and business rules. The work presented in [4] shows how process templates can be combined with business rules to design flexible business processes. This idea is applied to separate the basic process flow from the business policy elements. Another approach to capture variability in process models is represented in [27]. This approach proposes a formal framework for representing system variability that allows to detect circular dependencies and contradictory constraints in questionnaire models. In [7] an approach including formal representations and algorithms based on logical reasoning is proposed. Moreover, in this work, the validation in the context of customization of process variants is discussed.

Table 4 provides a comparative overview of the presented configuration approaches in light of our evaluation criteria (inspired from [28] and [29]): (i) Process Modeling Language, (ii) Formal Specification and (iii) Correctness Verification. The three symbols '+', '−' and '±' represent respectively that the

Table 4. Comparison table of related work and the proposed approach.

Approaches	Criteria		
	Process Modeling Language	Formal Specification	Correctness Verification
[22]	C-YAWL	+	±
[23]	CoSeNets	±	±
[1]	C-EPC	+	±
[24, 25]	C-EPC	+	+
[26]	annotated BPMN	±	−
[3]	any	±	+
[4]	Block-structured	±	+
[27]	C-EPC/ C-YAWL	±	±
[7]	Block-structured	+	+
[6]	C-BPMN	−	−
Proposed approach	C-BPMN	+	+

corresponding criteria is fulfilled, it is not fulfilled and it is partially fulfilled by the corresponding approach.

On the other hand, a number of works emphasize the value of SAT inside, for instance, product line engineering, business process, Cloud Computing, etc. For instance, in [11,12] propositional logic and SAT are used to analyze feature models which is a popular variability modeling notation used in product line engineering. [30] proposes the use of improved separation of duty algebra to describe a satisfiability problem of qualification requirements and quantification requirements. This is being done to provide a separation of duty and binding of duty requirements. And also in the other works [15,31], SAT-based approach is used to relax the failed queries through rewriting them in the Cloud Computing exactly in the Software as a Service (SaaS). In addition, SAT is adopted to compute a minimum composition within preserved-privacy of SaaS Services and Data as a Service (DaaS) Services for a given customer's request.

In summary, our approach combines the use of configurable business process with the SAT. The major differences from the above cited approaches are the following points: (1) It generates structurally correct process configuration options that do not contain run-time errors. (2) It generates all correct options from the beginning (at design time) which allows to assist process designer during the configuration time. (3) Since SAT has gained considerable audience with the advent of a new generation of SAT solvers during these last few years, the application of SAT techniques to configurable business process offers many benefits in terms of their analyzes concerning the generation of correct configurations.

6   Conclusion and Further Work

In this work, we have presented a formalization and correctness verification to provide all possible correct configurations in configurable business process. We propose a method to transform configurable business process to propositional satisfiability that will help improve their use and their adoption. We have showed the applicability of our approach and conducted experiments to show the efficiency of the approach.

As future work, we aim to work firstly on additional workflow elements, for example: resources, messages, etc. We also plan to study different constraints and requirements, for instance: temporal, business, QoS, etc., in order to exploit the integrated representation of the common and variable parts of a family of processes of business processes.

References

1. Rosemann, M., Van der Aalst, W.M.: A configurable reference modelling language. Inf. Syst. **32**(1), 1–23 (2007)
2. Recker, J., Rosemann, M., van der Aalst, W., Mendling, J.: On the syntax of reference model configuration – transforming the C-EPC into lawful EPC models. In: Bussler, C.J., Haller, A. (eds.) BPM 2005. LNCS, vol. 3812, pp. 497–511. Springer, Heidelberg (2006). https://doi.org/10.1007/11678564_46

3. Hallerbach, A., Bauer, T., Reichert, M.: Capturing variability in business process models: the Provop approach. J. Softw. Maint. Evol. Res. Pract. **22**(6–7), 519–546 (2010)
4. Kumar, A., Yao, W.: Design and management of flexible process variants using templates and rules. Comput. Ind. **63**(2), 112–130 (2012)
5. GröNer, G., BošKović, M., Parreiras, F.S., GašEvić, D.: Modeling and validation of business process families. Inf. Syst. **38**(5), 709–726 (2013)
6. Assy, N., Gaaloul, W.: Extracting configuration guidance models from business process repositories. In: Motahari-Nezhad, H.R., Recker, J., Weidlich, M. (eds.) BPM 2015. LNCS, vol. 9253, pp. 198–206. Springer, Cham (2015). https://doi.org/10.1007/978-3-319-23063-4_14
7. Asadi, M., Mohabbati, B., Gröner, G., Gasevic, D.: Development and validation of customized process models. J. Syst. Softw. **96**, 73–92 (2014)
8. La Rosa, M., van der Aalst, W.M., Dumas, M., Ter Hofstede, A.H.: Questionnaire-based variability modeling for system configuration. Softw. Syst. Model. **8**(2), 251–274 (2009)
9. van der Aalst, W.M., Dumas, M., Gottschalk, F., Ter Hofstede, A.H., La Rosa, M., Mendling, J.: Preserving correctness during business process model configuration. Formal Aspects Comput. **22**(3–4), 459–482 (2010)
10. Hallerbach, A., Bauer, T., Reichert, M.: Guaranteeing soundness of configurable process variants in Provop. In: 2009 IEEE Conference on Commerce and Enterprise Computing, pp. 98–105. IEEE (2009)
11. He, F., Gao, Y., Yin, L.: Efficient software product-line model checking using induction and a SAT solver. Front. Comput. Sci. **12**(2), 264–279 (2018)
12. Mendonca, M., Wąsowski, A., Czarnecki, K.: SAT-based analysis of feature models is easy. In: Proceedings of the 13th International Software Product Line Conference, pp. 231–240. Carnegie Mellon University (2009)
13. Xiang, Y., Zhou, Y., Zheng, Z., Li, M.: Configuring software product lines by combining many-objective optimization and SAT solvers. ACM Trans. Softw. Eng. Methodol. (TOSEM) **26**(4), 14 (2018)
14. Marques-Silva, J.P., Sakallah, K.A.: Boolean satisfiability in electronic design automation. In: Proceedings of the 37th Annual Design Automation Conference, pp. 675–680. ACM (2000)
15. Wakrime, A.A.: Satisfiability-based privacy-aware cloud computing. Comput. J. **60**, 1760–1769 (2017)
16. Assy, N.: Automated support of the variability in configurable process models. Ph.D. thesis, University of Paris-Saclay, France (2015)
17. Tseitin, G.: On the complexity of derivations in the propositional calculus. In: Slesenko, H. (ed.): Structures in Constructives Mathematics and Mathematical Logic, Part II, pp. 115–125 (1968)
18. Moskewicz, M.W., Madigan, C.F., Zhao, Y., Zhang, L., Malik, S.: Chaff: engineering an efficient SAT solver. In: Proceedings of the 38th Annual Design Automation Conference, pp. 530–535. ACM (2001)
19. Eén, N., Sörensson, N.: An extensible SAT-solver. In: Giunchiglia, E., Tacchella, A. (eds.) SAT 2003. LNCS, vol. 2919, pp. 502–518. Springer, Heidelberg (2004). https://doi.org/10.1007/978-3-540-24605-3_37
20. Zhang, L., Madigan, C.F., Moskewicz, M.H., Malik, S.: Efficient conflict driven learning in a Boolean satisfiability solver. In: Proceedings of the 2001 IEEE/ACM International Conference on Computer-Aided Design, pp. 279–285. IEEE Press (2001)

21. Kiepuszewski, B., ter Hofstede, A.H., van der Aalst, W.M.: Fundamentals of control flow in workflows. Acta Inform. **39**(3), 143–209 (2003)
22. Gottschalk, F., Van Der Aalst, W.M., Jansen-Vullers, M.H., La Rosa, M.: Configurable workflow models. Int. J. Coop. Inf. Syst. **17**(02), 177–221 (2008)
23. Schunselaar, D.M.M., Verbeek, E., van der Aalst, W.M.P., Raijers, H.A.: Creating Sound and reversible configurable process models using CoSeNets. In: Abramowicz, W., Kriksciuniene, D., Sakalauskas, V. (eds.) BIS 2012. LNBIP, vol. 117, pp. 24–35. Springer, Heidelberg (2012). https://doi.org/10.1007/978-3-642-30359-3_3
24. van der Aalst, W.M.P., Dumas, M., Gottschalk, F., ter Hofstede, A.H.M., La Rosa, M., Mendling, J.: Correctness-preserving configuration of business process models. In: Fiadeiro, J.L., Inverardi, P. (eds.) FASE 2008. LNCS, vol. 4961, pp. 46–61. Springer, Heidelberg (2008). https://doi.org/10.1007/978-3-540-78743-3_4
25. van der Aalst, W.M., Lohmann, N., La Rosa, M.: Ensuring correctness during process configuration via partner synthesis. Inf. Syst. **37**(6), 574–592 (2012)
26. Schnieders, A., Puhlmann, F.: Variability mechanisms in e-business process families. BIS **85**, 583–601 (2006)
27. La Rosa, M., Van Der Aalst, W., Dumas, M., ter Hofstede, A.: Questionnaire-based variability modeling for system configuration. Softw. Syst. Model. **8**(2), 251–274 (2008)
28. Rosa, M.L., Van Der Aalst, W.M., Dumas, M., Milani, F.P.: Business process variability modeling: a survey. ACM Comput. Surv. (CSUR) **50**(1), 2 (2017)
29. Boubaker, S.: Formal verification of business process configuration in the Cloud. PhD thesis, University of Paris-Saclay, France (2018)
30. Bo, Y., Xia, C., Zhang, Z., Lu, X.: On the satisfiability of authorization requirements in business process. Front. Comput. Sci. **11**(3), 528–540 (2017)
31. Ait Wakrime, A., Benbernou, S., Jabbour, S.: Relaxation based SaaS for Repairing Failed Queries over the Cloud Computing. In: 2015 IEEE 12th International Conference on e-Business Engineering (ICEBE). IEEE (2015)

Machine Learning and Data Analytics

TSWNN+: Check-in Prediction Based on Deep Learning and Factorization Machine

Chang Su, Ningning Liu, Xianzhong Xie$^{(\boxtimes)}$, and Shaowen Peng

College of Computer Science and Technology, Chongqing University of Posts
and Telecommunications, Chongqing 400000, China
{changsu, xiexzh}@cqupt.edu.cn,
1334696838@qq.com, 1769406307@qq.com

Abstract. With the increasing popularity of location-aware social media applications, Point of interest (POI) predictions and POI recommendations have been extensively studied. However, most of the existing research is to predict the next POI for a user. In this paper, we consider a new research problem, which is to predict the number of users visiting the POI during a particular time period. In this work, we extend the TSWNN model structure and propose a new method based on Factorization Machine (FM) and Deep Neural Network (DNN) to learn check-in features—TSWNN+. More specifically, this paper uses the Factorization Machine to learn the latent attributes of user check-in and time features. Then, the DNN is used to bridge time features, space features, and weather features to better mine the latent check-in behavior pattern of the user at the POI. In addition, we design a negative instance algorithm to augment training samples. In order to solve the problem of gradient disappearance caused by DNN, the residual structure is adopted. The experimental results on two classical LBSN datasets—Gowalla and Brightkite show the superior performance of the constructed model.

Keywords: Deep Learning · FM · Residual · Check-in prediction

1 Introduction

LBSN is a social network established through intelligent terminal devices. It is an online platform for users to share information such as hobbies, living conditions and tourism activities. It can truly reflect the user's social relationship and real-life activities, and connect the real physical world with the virtual network space. Social networking sites such as Brightkite, Gowalla have a large number of users. When users visit POI, they will check-in and post pictures, comments, strategies, etc. on the registered LBSN website. This forms a set of valuable POI registration data. These information truly records when the user visited the location. By learning the registration information, we can mine the potential rules and preferences of the user's behavior, and better predict the user' location. Location prediction based on LBSN is a relatively new research direction and can be widely applied in intelligent transportation [1], improving urban planning [2], tourism planning [3].

© Springer Nature Switzerland AG 2019
M. Younas et al. (Eds.): Innovate-Data 2019, CCIS 1054, pp. 65–79, 2019.
https://doi.org/10.1007/978-3-030-27355-2_5

The goal of this paper is to explore the user's movement regularity and predict whether users will check-in at a specific POI by analyzing the time features, space features and weather features corresponding to records when users check-in at POIs. Most of the existing studies focus on exploring from the users' perspective. In contrast, within the scope of what we know, there are few studies to predict the number of future visits at a POI [4]. This paper examines a new research problem from the POIs' perspective: Given a POI, predict the number of users visiting the POI during a particular time period. For instance, given a restaurant, it can be predicted how many customers may visit the restaurant, so that users can decide whether to visit or avoid overcrowding when choosing a restaurant they like. From this point of view, predicting potential user visits is very valuable. The predicted results can be applied to many scenarios, such as helping merchants predict the trend of user visit in the future, developing appropriate marketing plans, and timely launching appropriate activities to attract more users.

Check-in prediction faces great challenges. First of all, the user's behavior has periodicity [5]. For example, various fast-food restaurants are more likely to be visited by users during the daytime from Monday to Friday, while entertainment venues are more likely to be visited during the weekend. And weather factors also have an impact on user visit behavior. Therefore, the learning of time-space-weather features is necessary for prediction. Secondly, we can only observe explicit data in our dataset, that is, the location where the user has checked-in, but the dataset that only focuses on explicit data is usually sparse, and sparse data leads to low prediction accuracy. This requires integrating negative instances from implicit data into the dataset for consideration.

The main contributions of this paper are as follows:

- Integrating Factorization Machine (FM) and Deep Learning, a novel check-in prediction model (TSWNN+) is constructed, which fully explore and learn the user's check-in rules and patterns through the weather features, time features and space features to accurately predict potential users at a given POI.
- The space features and weather features are continuous. However, the time features are discrete. These features cannot be directly interacted. For discrete data, we use the embedded layer and FM map it to a low-dimensional sparse vector. Then use FM to remove sparsity, and finally use DNN to combine continuous features with discrete features.
- In order to solve the gradient disappearance problem caused by DNN, we adopt the residual structure.
- Aiming at the problem of user data sparsity, we propose an algorithm for constructing negative instances based on the information volume formula.

Experiments conducted on two classical LBSN datasets—Gowalla and Brightkite show the superior performance of TSWNN+ model.

2 Related Work

The prediction of POI based on LBSN is currently receiving widespread attention and extensive research. Most of the existing studies are based on user-centered, following the classical based on users collaborative filtering techniques that score POIs in terms of similarity between users' check-in activities [6, 7]. The techniques are further improved by taking into account social network and space influence in the prediction [8, 9]. Some researchers have focused their research on user history check-in data and proposed prediction algorithms for user check-in locations that rely on regular or repetitive modelling [10, 11]. Some studies have focused on using a set of related features to achieve superior predictive performance. For example, various hand-crafted features have been used in some studies [12, 13]. But these models lack the ability to automatically capture information from check-in data. To overcome this shortcoming, many studies have introduced Deep Learning models to automatically learn the best internal representation of features [14, 15]. DNNs are able to extract the hidden structures and intrinsic patterns at different levels of abstractions from training data. DNNs have been successfully applied in computer vision [16], speech recognition [17] and natural language processing (NLP) [18]. In the case where the input data is highly sparse, the traditional model has the challenge that the parameters are difficult to learn. In order to solve this problem. [19, 20] use a FM based on Matrix Factorization. FM can perform parameter learning well even in the case of sparse data.

[4] studies the prediction of a specific POI check-in in the future. This work proposes a Deep Learning model called TSWNN. The structure not only captures local time features, but also captures space features and weather features. However, the model has certain limitations. First, TSWNN directly feeds into the neural network after processing each check-in feature, and the features are not well connected. In addition, TSWNN ignores the problem that of gradient disappearance in DNNs. In this paper, We have taken advantage of TSWNN and made improvements based on it. In order to simulate user preferences, the traditional prediction system only focuses on explicit data [23], but the dataset that only focuses on explicit data is usually sparse, and the sparse data leads to low prediction accuracy. This requires the integration of negative instances from implicit data into the dataset for consideration [21]. In this paper, we design a negative instance sampling algorithm.

3 Feature Analysis and Feature Description

3.1 Feature Analysis

The data features used in this paper mainly include weather features (temperature, rainfall, wind speed), time features (month, day, hour, week), space features (longitude, latitude). These features include the user's check-in rules and habits. We can predict the user's future check-in behavior by learning these features.

Figure 1 shows the effect of dynamic changes of features on check-in in New York. Intuitively, the lower the wind speed, the more the number of check-ins, but this is not the case. It can be seen from Fig. 1(a) that not only the high wind speed corresponds to fewer check-ins, but the low wind speed corresponds to a small number of check-ins. Because high wind speeds are often accompanied by heavy rainfall, and low wind speeds often correspond to high temperatures. It can be seen from Fig. 1(b), as the rainfall increases, the number of users check-ins sharply decreases, which is also in line with people's daily habits. Generally, people are less likely to go out during the rain, and they tend to visit indoor places even if they go out. It can be seen from Fig. 1(c) that people tend to choose weather with suitable temperature for outside activities. Figure 1(d) shows that the user's check-in percentage on weekends is significantly higher than the weekday. The above conclusions also confirm the fact that these features include the user's check-in rules and patterns. We will analyze these features, learn the rules of each user, and then predict the number of future visits at a POI.

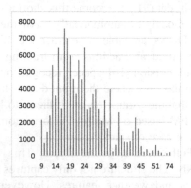
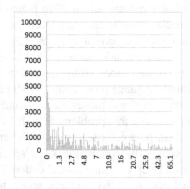

(a)Number of check-ins at different wind speeds (b) Number of check-ins under different rainfall

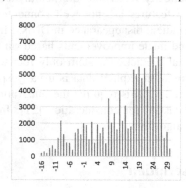
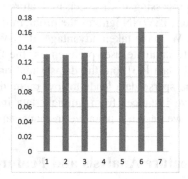

(c)Number of check-ins at different temperatures   (d)One-week check-in ratio

Fig. 1. Effect of dynamic changes of features on check-in

3.2   Feature Description

Time Feature. According to common sense, the user's visit behavior generally changes over time. Users will visit different POIs in different time periods. In other words, there is an implicit relationship between time features, such a relationship reflects the user's behavior pattern. Therefore, our check-in time is accurate to the hour, that is, <month, day, hour, week>, which are discrete features. For example, when a user goes to Starbucks at time $t, t \in \{0, 1, 2.., 23\}$. These discrete time features are usually expressed in the form of multi-field categorical, it is usually converted to high-dimensional sparse binary features by one-hot encoding. For instance, [Month = May] with one-hot encoding can be expressed as <0, 0, 0, 0, 1, 0, 0, 0, 0, 0, 0, 0>. How to learn the behavior pattern of users from sparse vectors is a difficult problem. This paper will use the idea of FM to process high-dimensional sparse data. After the FM structure processing, the second-order interaction between the features can be learned, and the sparseness problem of the data is solved.

Space Feature. At first, space information have fundamental effects in revealing characteristics of the user and are helpful for the behavior modeling, which are reflected in many studies [14, 15]. In addition, the space features can accurately describe the geographic location of POIs. In daily life we often use descriptions such as 'beside the department store' However, such a description cannot accurately express the location. The space features need to accurately and unambiguously describe the location. Therefore, this paper uses latitude and longitude to describe the geographic information. It can accurately locate POI through latitude and longitude, and it is more convenient for model to deal with space features.

Weather Feature. Considering weather information is an important factor in check-in prediction, it is necessary to involve it into our model. The dynamic changes and unpredictable factors of the weather will affect the periodicity of the user's check-in. Therefore, we introduce weather data and use it as continuous features such as temperature, rainfall, and wind speed. By learning these features, we can better discover the user's check-in rules.

4 Proposed Model

4.1 Model Description

The architecture of the TSWNN+ model is illustrated in Fig. 2 from a down-top perspective, where W_x, b_x denote the corresponding weight matrix and bias vector in the model, respectively.

(1) As introduced in Sect. 3.2, time features are discrete features data which are categorical and contain multiple fields. The time features are transformed into the high-dimensional sparse vector via one-hot encoding. However, the high-dimensional sparse data cannot be directly feed into the deep neural networks. On the other hand, interactive patterns between interfiled categories are importance to learn. Hence, an embedding layer is set up to transform the discrete features into a low-dimensional space and explore feature interactions. First, time features are the input as discrete multi-field feature $Field[1..i..N]$. Then the discrete features are embedded to the low-dimensional vectors $T[1..i..N]$, where the length of T_i is M:

$$T_i = W_0^i Field_i \qquad (1)$$

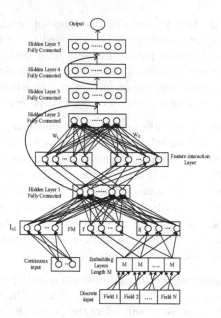

Fig. 2. TSWNN+

(2) FM was proposed in [20]. It is a general predictor that can capture feature interactions more efficiently than traditional methods when data is sparse. As shown in Eq. (2):

$$\hat{y}_{FM} = w_0 + \sum_{i=1}^{N} w_i x_i + \sum_{i=1}^{N} \sum_{j=i+1}^{N} <T_i, T_j> x_i x_j \tag{2}$$

Where $w_0 \in R, w \in R_N, V \in R^{N \times K}$ are the parameters to be estimate; $<\cdot,\cdot>$ is the dot product of two factor vectors of size K. We elaborate the mechanism of FM using a example, if there are check-in records for user A at 2 pm on Wednesday and 2 pm on Saturday, then $<v_{14}, v_{Wed}>$ and $<v_{14}, v_{Sat}>$ are similar. The latent vectors v_{14}, v_{Wed}, v_{Sat} would be learnt. Similarly, if there is check-in record at 3 pm on Wednesday, then the latent vector $<v_{15}, v_{Wed}>$ would be updated. Therefore, $<v_{15}, v_{Sat}>$ would be calculated and the possibility that user A checks in at 15 pm on Saturday would be estimated. This means that even with only one interaction of the data also helps to estimate the parameters of the relevant interactions. Hence, we use the network structure of FM to process sparse data, where f is the first-order relationship of the features, and S is the second-order relationship between features. After the processing of this structure, we learn the second-order relationship and eliminate sparsity. Here, each feature is a vector T_i, and the product between the features is the dot product of the vectors:

$$f = f[1..i..N] \tag{3}$$

$$f_i = W_f^i T_i \tag{4}$$

$$S = \{s_{i,j}\}(i,j = 1...N) \tag{5}$$

$$s_{i,j} = <R_i R_j> <T_i T_j> \tag{6}$$

(3) For continuous features (weather information, space information), this paper adds a fully-connected hidden layer for learning, and defines $relu(x) = \max(0, x)$ as the excitation function of the hidden layer output:

$$l_{c1} = relu(W_{c1}x + b_{c1}) \tag{7}$$

$$l_{c2} = W_{c2}l_{c1} + b_{c2} \tag{8}$$

(4) l_1 is the output of hidden layer h_1, whose input consists of the processing results of discrete features and continuous features:

$$l_f = W_f' f \tag{9}$$

$$l_S = W_S' S \tag{10}$$

$$l_1 = relu(l_{c2} + l_f + l_S + b_0) \tag{11}$$

(5) l_2 is the output of hidden layer h_2. At this time, the discrete features data have been processed by the FM layer, Then the data enters the hidden layer h_2 through the first-order linear relationship and second-order feature interaction structure. Where $\otimes$ represents the multiplication of two vectors of the same size, as shown in Eq. (12). The potential link between continuous features and discrete features can be better mined through this structure. Thereby, the prediction accuracy is improved.

$$l_2 = relu\{W_L W_l \otimes l_1 + W_Q(W_q W_q') \otimes (l_1^t l_1) + b_1\} \tag{12}$$

(6) In fact, using deep neural networks will face the problem of gradient disappearance, because the deeper the number of layers, the more parameters are in the network. It is more difficult to learn the parameters of hidden layer neurons near the input layer. Therefore, we use the residual structure [22] to solve the problem of gradient disappearance. l_3 is the output of hidden layer h_3. The input of h_3 comes from l_1 and l_2 due to the residual structure. Therefore, the stability of the gradient during parameter learning is guaranteed. Similarly, in order to improve the accuracy, add h_4, h_5, at the same time, the input of h_5 comes from l_3 and l_4.

Finally, the output of TSWNN+ is a real number $\hat{y} \in (0, 1)$ as the result of the prediction:

$$l_3 = relu(W_2 l_2 + l_1 + b_2) \tag{13}$$

$$l_4 = relu(W_3 l_3 + b_3) \tag{14}$$

$$l_5 = relu(W_4 l_4 + l_3 + b_4) \tag{15}$$

$$\hat{y} = sigmoid(W_5 l_5 + b_5) \tag{16}$$

4.2 Loss Function

In this paper, we use binary cross-entropy loss as a loss function, aiming to minimize the loss function to achieve the goal of maximizing correct prediction. Its optimization can be done by performing SGD (Stochastic Gradient Descent):

$$Loss = -[y \log \hat{y} + (1 - y) \log(1 - \hat{y})] \tag{17}$$

y indicates whether the user is checked-in at the location, and $\hat{y}$ represents the predicted result.

4.3 Implicit Data Expansion

The method we proposed is a binary classification task. We define a label: y_{ij}, when $y_{ij} = 1$ indicates that the user has checked-in at the POI, and $y_{ij} = 0$ indicates that the user has not checked-in at the POI. However, in the dataset we can only observe the location where the user has an interaction relationship, that is, where the user has checked-in. Therefore, we design a sampling algorithm for negative samples. That is, construct the data of $y_{ij} = 0$. The training model uses two types of data. The first type is the check-in data shows where the user has visited (classified as 1). The second type of data is to sample negative instances from the dataset. (classified as 0).

The construction of negative instances can expand the dataset to a certain extent and alleviate the problem of data sparsity, thus improving the prediction accuracy. As described in Table 1. We use the information volume formula to construct the implicit data required. The information volume formula is shown in (18):

$$\delta = -\log p_i \tag{18}$$

For example, if user 1 goes to POI A at time t, then the probability that user 1 visits POI A at time t is 1, that is $P_A^t = 1$. User 1 must not be able to check-in at the POI B at time t. Therefore, the amount of information that user 1 does not visit the POI B at time t is 0 by Eq. (19). Such instances are redundant for constructing implicit data. Therefore, the constructed implicit data must satisfy the user's absence of the check-in record at that time (as described in Algorithm 1). According to this construction

Table 1. Implicit data expansion algorithm

Algorithm 1: Negative instance construction

Input: User's check-in dataset.

Output: Negative Instances.

1 . For all users:

2 . i <--0;

3 . Do

4 .   Check_nums<--number of check-in times for a user;

5 .   Create_nums=n*Check_nums; #Create n times check-in records

6 .   For Create_nums:

7 .   Random selection of a check-in record;

8 .   If the user does not have a check-in record at the same time:

9 .   Then construct this record as a negative instance;

10 .   else:

11 .   Re-extract the next instance;

12 .   i++;

13 . Until dataset is empty;

14 . Return negative instances;

---

method, we can expand the dataset by the ratio of 1:n for positive instances and negative instances.

$$P_A^t = 1$$
$$I = - \log \bar{p}_B^t = 0 \qquad (19)$$

In the process of negative instances collection, it is highly probable that the locations that are not accessible to the user are incorrectly selected as negative instances when the negative instance dataset is too large. These misselected instances reduce the quality of the negative instance set and the quality of the training sample set, which in turn affects the accuracy of the model. A large number of experimental results show that the best results are obtained when the ratio of positive instances and negative instances is 1:4.

5 Experiments and Results

5.1 Dataset

To validate whether the TSWNN+ model can really help enhance the check-in prediction accuracy, we performed a series of experiments using two real-world datasets Gowalla and Brightkite. We filter out the users with fewer than ten check-ins at the POIs. Detailed statistics of the dataset can be found in Table 2. In addition, we select the POIs with a large number of check-ins as the predicted point, as shown in Table 3.

Table 2. User's check-in in the datasets

Datasets	Cities	Users	Number of check-ins
Gowalla	New York	2338	113094
Gowalla	Los Angeles	1019	39470
Gowalla	Chicago	1044	49192
Brightkite	New York	1060	80353
Brightkite	Los Angeles	576	54183
Brightkite	Chicago	499	41585

Table 3. POI with a large number of check-ins in the datasets

Datasets	Cities	Longitude	Latitude	Number of check-ins	Location
Gowalla	New York	−73.87200594	40.77457811	1141	La Guardia Airport
Gowalla	Los Angeles	−118.4071684	33.94353796	1387	Los Angeles International Airport
Gowalla	Chicago	−87.90367126	41.97802863	1269	O'Hare International Airport
Brightkite	New York	−74.005973	40.714269	11920	Municipal government office
Brightkite	Los Angeles	−118.243685	34.052234	12058	Columbia Television Studio
Brightkite	Chicago	−87.650052	41.850033	5623	Express delivery station

5.2   Evaluation Procedure and Performance Metric

Accuracy and Area Under ROC Curve (AUC) are two commonly used global evaluation indicators for prediction. In order to measure the overall performance of the model, we use Accuracy and AUC as indicators to objectively reflect the prediction effect. The larger the value, the better the performance. The AUC is one of the main evaluation indicators of the binary classification model [14]. The accuracy is the ratio of the correct prediction of the model, as shown in Eq. (20).

$$Accuracy = \frac{TP + TN}{TP + FP + TN + FN} \tag{20}$$

5.3   Baselines

The research direction of the check-in prediction problem is quite novel. At present, the research on this direction is rarely seen. Therefore, in order to investigate the model effectiveness, we have compared TSWNN+ with several representative methods:

(1) TSWNN−: When TSWNN+ without residual structure.
(2) TSWNN: This is the most popular model for check-in prediction proposed in [4].
(3) NN: A common neural network model with two hidden layers.
(4) FM: Discrete data is processed by FM [20], and output by Logistic Regression.
(5) Support Vector Machine (SVM): Use the sigmoid function as a kernel function.

Parameter Settings. To determine hyperparameters of TSWNN+, we learned by optimizing the binary cross-entropy loss of Eq. (17). We have selected the optimal parameters through many experiments, where we sampled four negative instances per positive instance, M and K are set to 20, M is the length of the discrete features mapped to the low-dimensional vector mentioned earlier in this paper, K is the columns of the coefficient matrix $(N \times K)$ of the second-order relationship in the FM. Initial learning rate α was set to 0.05, which is linearly attenuated during the learning process and is conducive to model learning.

5.4   Results and Analysis

The performance comparison on the two datasets evaluated by Accuracy and AUC is illustrated in Fig. 3.

SVM shows poor performance on the datasets. FM slightly improves the results comparing with SVM, but does not predict well the number of users visiting the POI during a particular time period. And the accuracy of NN prediction has been improved comparing with FM and SVM. This shows that Deep Learning can improve prediction performance. Our model TSWNN+ combines the advantages of both FM and Deep Learning. TSWNN− adds multiple hidden layers based on the TSWNN model but without residual structure. We can see that the Accuracy and AUC of the predictions have not increased compared to TSWNN, but have dropped sharply. The reason is that the deeper the number of layers, the more parameters are in the network. It is more

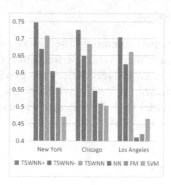

(a) Accuracy on Gowalla

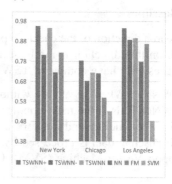

(b) AUC on Gowalla

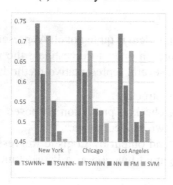

(c) Accuracy on Brightkite

(d)AUC on Brightkite

Fig. 3. Accuracy and AUC on datasets

difficult to learn the parameters of hidden layer neurons near the input layer. Thus, the phenomenon of gradient disappearing appears. TSWNN+ avoids the gradient disappearing due to the residual structure. No matter how deep the neural network is, the stable learning rate can be obtained. Table 4 shows the excellent performance of TSWNN+ in predicting results. Compared with TSWNN and TSWNN−, the accuracy is improved by 5.1% and 13.1%, respectively, while the AUC is improved to 3.5% and 8.7%, respectively. These great improvements indicate that our constructed TSWNN+ can better model time features, space features and weather features.

Table 4. Comparison of models

Model	Average accuracy	Decrease ratio	Average AUC	Decrease ratio
TSWNN+	72.4%	0%	90.6%	0%
TSWNN−	62.9%	13.1%	82.7%	8.7%
TSWNN	68.7%	5.1%	87.4%	3.5%
NN	52.4%	27.6%	70.8%	21.9%
FM	50.3%	30.5%	82.1%	9.4%
SVM	47.9%	33.8%	48.8%	46.1%

5.5   Performance Improvement

As mentioned in the data enhancement concept mentioned above, increasing the training set can improve the generalization performance of the test set. To illustrate the impact of negative sampling on prediction accuracy, we show the performance of TSWNN+ model at different negative sampling ratios. The abscissa is the ratio of negative instances to positive instances. The experimental results are shown in Fig. 4:

It can be clearly seen from experimental results that just one negative sample per positive instance is insufficient to achieve better performance, and it is beneficial to sample more negative instances. From Fig. 4(b), (d), the AUC value remains relatively stable during the increase of the data negative instances ratio. According to Fig. 4(a), (c), the accuracy rate has been greatly improved. When the ratio of positive and negative instances reaches 1:4, the accuracy rate has been increased to over 90%. This indicates that the increase of training data can improve the prediction effect, and greatly alleviate the problem of poor prediction caused by sparse datasets.

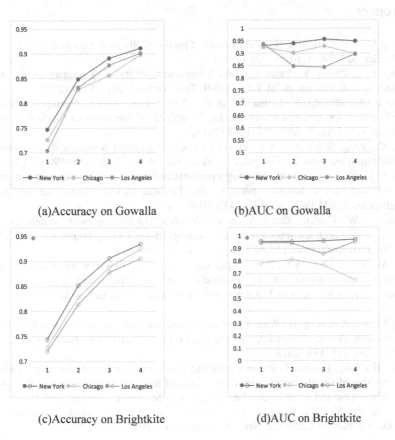

(a)Accuracy on Gowalla   (b)AUC on Gowalla

(c)Accuracy on Brightkite   (d)AUC on Brightkite

Fig. 4. Accuracy and AUC in the different ratio of negative to positive instances

6 Conclusion

This paper constructs a novel check-in prediction model (TSWNN+), which fully explore and learn the user's check-in rules and patterns through the weather information, time information and space information to accurately predict potential users at a given POI. This result can be applied to computer interdisciplinary such as tourism planning, public safety, etc. People can choose to go or not based on the predicted results of the tourist attractions. This will avoid overcrowding and provide a more comprehensive service for tourists. The experimental results show the superior performance of the constructed model.

Acknowledgments. This work was supported by the National Nature Science Foundation of China (grant numbers 61271259), the Chongqing Nature Science Foundation (grant numbers CSTC2016jcyjA0398, CTSC2012jjA40038), and the Research Project of Chongqing Education Commission (grant numbers KJ120501C).

References

1. Song, C., Qu, Z., Blumn, N., et al.: Limits of predictability in human mobility. Am. Assoc. Adv. Sci. **327**(5968), 1018–1021 (2010)
2. Tang, L.A., Zheng, Y., Yuan, J., et al.: A framework of traveling companion discovery on trajectory data streams. ACM Trans. Intell. Syst. Technol. **5**(1), 1–34 (2014)
3. Noulas, A., Scellato, S., Lathia, N., et al.: Mining user mobility features for next place prediction in location-based services. In: Twelfth IEEE International Conference on Data Mining, vol. 5, no. 1, pp. 1038–1043 (2013)
4. Su, C., Peng, S.W., Xie, X.Z., et al.: Check-in prediction based on deep learning and factorization machine. In: Computer Science , vol. 46, no. 5, pp. 185–190 (2019)
5. He, J., Li, X., Liao, L., et al.: Inferring a personalized next point-of-interest recommendation model with latent behavior patterns. In: Thirtieth AAAI Conference on Artificial Intelligence AAAI Press, pp. 137–143 (2016)
6. Zheng, V.W., Cao, B., Zheng, Y., et al.: Collaborative filtering meets mobile recommendation: a user-centered approach. In: Proceedings of the Twenty-Fourth AAAI Conference on Artificial Intelligence (AAAI), pp. 236–241 (2010)
7. Ye, M., Yin, P., Lee, W.C., et al.: Exploiting geographical influence for collaborative point-of-interest recommendation. In: Proceedings of the 34th International ACM SIGIR Conference on Research and Development in Information Retrieval (SIGIR), pp. 325–334 (2010)
8. Ye, J., Zhu, Z., Cheng, H.: What's your next move: user activity prediction in location-based social networks. In: Proceedings of the SIAM International Conference on Data Mining (SDM), pp. 171–179 (2013)
9. Gao, H., Tang, J., Liu, H.: gSCorr: modeling geo-social correlations for new check-ins on location-based social networks. In: Proceedings of the 21st ACM International Conference on Information and Knowledge Management (CIKM), pp. 1582–1586 (2012)
10. Cao, J., Xu, S., Zhu, X., et al.: Effective fine-grained location prediction based on user check-in pattern in LBSNs. J. Netw. Comput. Appl. **108**, 64–75 (2018)

11. Chang, J., Sun, E.: Location3: how users share and respond to location-based data on social. In: International Conference on Weblogs and Social Media, Barcelona, Catalonia, Spain, pp. 74–80. DBLP, July 2011

12. Long, V., Nguyen, P., Nahrstedt, K., et al.: Characterizing and modeling people movement from mobile phone sensing traces. Pervasive Mob. Comput. **17**, 220–235 (2015)

13. Do, T.M.T., Gatica-Perez, D.: Where and what: using smart phones to predict next locations and applications in daily life. Pervasive Mob. Comput. **12**, 79–91 (2014)

14. Liu, Q., Wu, S., Wang, L., et al.: Predicting the next location: a recurrent model with spatial and temporal contexts. In: Thirtieth AAAI Conference on Artificial Intelligence AAAI Press, pp. 194–200 (2016)

15. Al-Molegi, A., Jabreel, M., Ghaleb, B.: STF-RNN: space time features-based recurrent neural network for predicting people next location. In: IEEE Symposium Series on Computational Intelligence, pp. 1–7 (2016)

16. Zeiler, M.D., Taylor, G.W., Fergus, R.: Adaptive deconvolutional networks for mid and high level feature learning. In: International Conference on Computer Vision IEEE Computer Society, pp. 2018–2025 (2011)

17. Li, D., Abdelhamid, O., Yu, D.: A deep convolutional neural network using heterogeneous pooling for trading acoustic invariance with phonetic confusion. In: IEEE International Conference on Acoustics, pp. 6669–6673 (2013)

18. She, Y., He, X., Gao, J., et al.: A latent semantic model with convolutional-pooling structure for information retrieval, pp. 101–110. ACM (2014)

19. Guo, H., Tang, R., Ye, Y., et al.: DeepFM: a factorization-machine based neural network for CTR prediction. In: The Twenty-Sixth International Joint Conference on Artificial Intelligence, vol. 1703, pp. 1725–1731 (2017)

20. Rendle, S.: Factorization machines. In: IEEE International Conference on Data Mining, pp. 995–1000 (2010)

21. He, X.N., Liao, L.Z., Zhang, H.W., et al.: Neural collaborative filtering. In: Proceedings of the 26th International Conference on World Wide Web, pp. 173–182 (2017)

22. He, K., Zhang, X., Ren, S., et al.: Deep residual learning for image recognition. In: IEEE Conference on Computer Vision and Pattern Recognition, pp. 770–778 (2015)

23. Zheng, Y., Tang, B., Ding, W., et al.: A neural autoregressive approach to collaborative filtering. In: Proceedings of the Thirty-Third International Conference on Machine Learning, pp. 764–773 (2016)

Deep Learning Based Sentiment Analysis on Product Reviews on Twitter

Aytuğ Onan[✉]

Faculty of Engineering and Architecture, Department of Computer Engineering,
Izmir Katip Çelebi University, 35620 Izmir, Turkey
aytug.onan@ikc.edu.tr

Abstract. Sentiment analysis is the process of extracting an opinion about a particular subject from text documents. The immense quantity of text documents contain opinions or reviews towards a particular entity. The identification of sentiment can be useful for individual decision makers, business organizations and governments. Sentiment analysis is an important research direction. Deep learning is a recent research direction in machine learning, which builds learning models based on multiple layers of representations and features of data. Deep learning based frameworks can be employed in a wide range of applications, including natural language processing tasks, with encouraging prediction results. In this paper, we present a deep learning based scheme for sentiment analysis on Twitter messages. In the presented scheme, three-word embeddings based schemes (namely, GloVe, fastText and word2vec) and convolutional neural network (CNN) have been utilized. In the empirical analysis, different subsets of Twitter messages, ranging from 5000 to 50.000 are taken into consideration. The prediction results obtained by deep-learning based schemes have been compared to conventional classifiers (such as, Naïve Bayes and support vector machines).

Keywords: Sentiment analysis · Deep learning · Word embeddings · Word2vec

1 Introduction

Sentiment analysis, also known as, opinion mining is a subfield of natural language processing, dedicated to extracting subjective information encountered in text documents, such as opinions, sentiments, attitudes and evaluations. Sentiment analysis involves several tasks, such as polarity detection (positive or negative), stance detection (identifying relative stance/perspective of text documents) and aspect identification towards a target entity [1–3]. With the advances in social media, the immense quantity of user-generated unstructured text, containing opinions and sentiments become available on the Web. Social media users can share and/or follow public opinions towards an entity, an event or a service on microblogging platforms. This information can be valuable to business organizations, governments and individual decision makers [4]. The identification of public sentiments towards policies, products and organizations can be very beneficial to organizations and it can be utilized to serve decision

© Springer Nature Switzerland AG 2019
M. Younas et al. (Eds.): Innovate-Data 2019, CCIS 1054, pp. 80–91, 2019.
https://doi.org/10.1007/978-3-030-27355-2_6

support systems and individual decision makers [5]. As a consequence, the analysis of text documents available on social media platforms is essential task and sentiment analysis is a promising research direction.

Based on the level of granularities, sentiment analysis can be conducted on three-levels, namely, there are document-level, sentence-level and entity/aspect level schemes to sentiment classification [6]. Document-level schemes seek to identify the polarity of text documents based on a single entity/or product. Sentence-level schemes seek to identity the polarity and sentiment orientation of subjective sentences. In addition, aspect-level schemes seek to identify sentiments by focusing on particular features/aspects of entities [7].

Sentiment analysis methods can be broadly divided into two categories, as lexicon based methods and machine learning based methods [8]. The lexicon-based schemes to sentiment analysis identify orientation of a text document by computing semantic orientation of words and phrases [1].

The lexicon-based schemes involve a dictionary of positive and negative sentiment values corresponding to words. The lexicon-based methods can be effectively solve sentiment analysis problems with scalability [9]. The machine learning based schemes to sentiment analysis are supervised learning tasks, where labelled text documents have been utilized. Supervised learning algorithms, such as support vector machines, Naïve Bayes, k-nearest neighbor and random forest have been successfully utilized for polarity detection [10].

Deep learning is a recent research direction of machine learning that aim to obtain classification models with high predictive performance based on multiple layers or stages of nonlinear information processing and supervised/or unsupervised learning feature representations in a hierarchical manner [11]. In the hierarchies of levels, higher levels have more distributed/compact representations toward the data compared to the lower levels of the architecture. In this way, complex relationships among lower-levels can be utilized. Deep learning is an interdisciplinary research field, which involves artificial intelligence, pattern recognition, signal processing and artificial neural networks. Deep learning can be utilized in a wide range of applications, including computer vision, speech recognition and natural language processing tasks, with encouraging predictive performance.

In this paper, we present a deep learning based approach to sentiment analysis on Twitter messages. In the presented scheme, three-word embeddings based schemes (namely, GloVe, fastText and word2vec) and convolutional neural network (CNN) have been utilized. In the empirical analysis, different subsets of Twitter messages, ranging from 5000 to 50,000 are taken into consideration. The prediction results obtained by deep-learning based schemes have been compared to conventional classifiers (such as, Naïve Bayes and support vector machines).

The rest of this paper is structured as follows: In Sect. 2, related work on sentiment analysis has been presented. Section 3 presents the methodology of the study, Sect. 4 presents the experimental procedure and empirical results of the study. Finally, Sect. 5 presents the concluding remarks.

2 Related Work

This section briefly reviews the existing works on deep learning-based schemes on sentiment analysis.

Glorot et al. [12] presented a deep learning framework based on Stacked Denoising Autoencoder with sparse rectifier units, which performs an unsupervised feature extraction for domain adaptation task of sentiment classification. In another study, dos Santos and Gatti [13] presented a deep learning-based architecture for sentiment classification on Twitter messages that jointly employs character-level, world-level and sentiment-level representations. In the presented scheme, convolutional neural networks have been utilized to identify character to sentence-level features. Twitter sentiment classification based on deep learning was the scope of another study conducted by Tang et al. [14]. The presented scheme consists of two feature representations, namely, linguistic features (such as, the number of words with all characters in uppercase, emoticons, elongated units, sentiment lexicon, negation, punctuation, and N-grams) and sentiment-specific word embedding based feature representations have been taken into consideration. Similarly, Severyn and Moschitti [15] introduced a convolutional neural network based approach for sentiment classification on Twitter. The presented scheme was applied to phrase-level and message-level sentiment analysis tasks.

In another study, Hu et al. [16] presented a deep learning based framework for sentiment analysis. In the presented scheme, linguistic and domain knowledge have been utilized in order to obtain feature representation. Then, a deep neural network has been applied on three sentiment analysis datasets from different domains, such as electronic product reviews, movie reviews and hotel reviews. In another study, Johnson and Zhang introduced a convolutional neural network based scheme for text categorization [17]. The presented scheme utilizes bag-of-word conversion based representation in the convolution layer.

In another study, Tang et al. [18] presented a deep learning based framework for sentiment classification. In the presented scheme, convolutional neural network has been utilized in order to learn sentence representations. Then, gated recurrent neural network has been utilized in order to identify semantics of sentences and their relationships. Similarly, Tang et al. [19] presented a document level sentiment classification scheme based on deep neural networks which incorporates user and product level information. Ruder et al. [20] utilized a convolutional neural network based architecture for aspect-level sentiment classification, with an application to several domains, including hotel, laptop, phone and camera reviews.

In a recent study, Sun et al. [21] introduced a deep belief network based scheme for sentiment analysis on microblogging platforms in Chinese. The presented scheme eliminates the feature sparsity problem encountered in short text documents. Similarly, Araque et al. [22] presented a deep learning sentiment classification scheme based on word embeddings based representation and linear machine learning classifier. In another study, Paredes-Valverde et al. [23] introduced a deep learning based approach based on word2vec feature vectors and convolutional neural network for sentiment analysis of Twitter messages in Spanish.

3   Methodology

This section presents the methodology of the study, namely, dataset utilized in the empirical analysis and preprocessing stages, word embedding based feature representation schemes and convolutional neural network architecture have been briefly presented. The deep learning based sentiment classification framework utilized in this study consists of three main components, as preprocessing module, word-embeddings based feature representation module and convolutional neural network module. After dataset collection, it has been preprocessed by applying tokenization and normalization methods. Then, word2vec, fastText and GloVe vector based feature representation schemes have been utilized in order to obtain feature vectors. Finally, sentiment classification on Twitter messages has been conducted by training the convolutional neural network. The details regarding the main modules of the presented scheme has been discussed in the following sections.

3.1   Dataset Collection and Preprocessing

In the dataset collection we have adopted the framework presented in [24, 25]. To build sentiment dataset on product reviews, we have determined a number of keyword related to technological products. We have collected approximately 62.000 tweets with various topics related to technological products, written in English. In order to collect dataset, Twitter4J, an open-source Java library for utilizing Twitter Streaming API, has been utilized. After obtaining Twitter messages, automatic filtering has been employed in order to remove duplicated tweets, retweets, ambiguous, irrelevant and redundant tweets. Each tweet is labelled manually by a single class label, as either positive or negative, to indicate the sentiment orientation of the review text. In this way, we obtained a collection of roughly 37.000 positive tweets and roughly 25.000 negative tweets. In order to obtain a balanced corpus, our final corpus contains a collection of review text with 25.000 positive and 25.000 negative reviews. In Table 1, the descriptive information regarding the distribution of dataset has been presented.

Table 1.   Descriptive information regarding the corpus utilized in empirical analysis

Set	Positive	Negative	Total number
Training Set	20.000	20.000	40.000
Testing Set	5.000	5.000	10.000

Twitter consists of messages with unstructured and irregular nature. As a consequence, preprocessing is an essential task to sentiment analysis on Twitter. In order to preprocess our corpus, we have adopted the framework presented in [23]. Initially, tokenization has been employed on the corpus, so that, review messages have been divided into tokens, namely, words or punctuation marks. The tokenization process has been conducted using Twokenize tool. After tokenization process, all the non-informative items generated by Twokenize has been eliminated. Then, mentions and replies to other users' tweets were eliminated, URLs were removed and special characters (such as "#" character) were eliminated.

3.2   Word Embeddings Based Feature Representation

In order to build deep learning based classification models on natural language tasks, one key issue is to represent text documents. Word embeddings based representation is an important scheme for language modelling and feature learning, in which words and documents are transformed into a vector of real numbers. Word embeddings have been successfully employed in text categorization and sentiment analysis, owing to their ability to extract syntactic and semantic relations among the words [26]. In this study, three well-known deep learning based word embedding schemes, namely, word2vec, global vectors (GloVe) and fastText have been taken into consideration. The rest of this subsection briefly reviews these schemes.

The word2vec is an unsupervised, computationally efficient, prediction model that learns word embeddings from text documents. The word2vec model consists of two models, namely, continuous bag of words model (CBOW) and continuous skip-gram model [27]. CBOW model predicts the center word from its context words, whereas skip-gram model predicts the particular context words based on the center word. Continuous bag of words model treats the entire context as one observation, whereas skip-gram model treats regards context-target pairs as observations [28]. Continuous bag of words model can yield promising results for small datasets, yet skip-gram model yields better predictive performance on larger datasets. Let we denote a sequence of training words $w_1, w_2, \ldots, w_T$ with length T, the objective of skip-gram model is determined based on Eq. 1 [29]:

$$argmax_\theta \frac{1}{T} \sum_{t=1}^{T} \sum_{-C \leq j \leq C, j \neq 0} log P_\theta \left(w_{t+j} | w_t \right) \tag{1}$$

where C denote the size of training context, $P\left(w_{t+j}|w_t\right)$ represents a neural network with a set of parameters denoted by θ.

The fastText word embedding scheme is an extension of word2vec scheme, which regards each word as a bag of character n-grams [30]. In comparison to word2vec scheme, fastText scheme can yield more robust word embeddings for rare words. Since fastText model takes character n-grams into account, good embedding schemes can be still obtained for rare words. In addition, the fastText scheme can be computationally efficient.

The global vectors (GloVe) is a global log-bilinear regression model for word embeddings based on global matrix factorization and local context window methods [31]. The objective of Glove model is determined based on Eq. 2:

$$J = \sum_{i,j=1}^{V} f\left(X_{ij}\right) \quad \left(w_i^T \omega_j + b_i + b_j - log X_{ij}\right)^2 \tag{2}$$

where V denotes the vocabulary size, $w \in R^d$ represent word vectors, $\omega \in R^d$ represent context word vectors, X denote co-occurrence matrix and X_{ij} denotes the number of times word j occurs in the context of word i. $f\left(X_{ij}\right)$ denotes a weighting function and b_i, b_j are bias parameters [31].

3.3 Convolutional Neural Network Architecture

Convolutional neural networks (CNNs) are a type of deep neural networks that yield promising results in a wide range of application fields, including image recognition and natural language processing [17, 32].

Convolutional neural networks process data with a grid-based topology. In contrast to conventional neural network structures, where matrix multiplication has been applied on the inner layers, convolutional neural networks has been characterized by convolution operation in their layers. Convolutional neural network consists of input layer, output layer and hidden layers. The hidden layers of the network consists of convolutional layer, pooling layers, fully connected layers and normalization layers. Convolutional layers apply convolution operation on the input data so that feature maps have been obtained. In addition to that, activation function has been utilized in conjunction with feature maps on convolutional layers. In this way, the nonlinearity has been added to the architecture [33]. The rectified linear unit is a typical activation function utilized on CNN. After convolution, pooling layers combine the outputs of neuron clusters. The typical pooling function is max pooling scheme, where the maximum value from each cluster has been taken. Pooling layers reduce the number of instances in each feature map and reduce the training time. After convolutional and pooling layers, fully connected layers determine the final output of the architecture [34].

4 Experimental Procedure and Results

In this section, experimental procedure, evaluation measures and results of the study has been presented.

4.1 Experimental Procedure

In the empirical analysis, the predictive performance of convolutional neural networks has been compared to conventional classifiers, such as Naïve Bayes and support vector machines. For conventional classifiers, the default parameters of the algorithms have been utilized in the empirical analysis. In addition to that, different subsets of Twitter messages, ranging from 5000 to 50.000 are taken into consideration. For each subset of the corpus, 80% of data has been utilized as the training set, whereas the rest of data has been utilized as the testing set. In order to label Twitter messages as positive and negative tweets, TensorFlow implementation of convolutional neural networks has been utilized.

4.2 Evaluation Measures

In order to evaluate the predictive performance of conventional classification algorithms and deep-learning based schemes on sentiment analysis, we have utilized F-measure, as the evaluation measure.

Precision (PRE) is the proportion of the true positives against the true positives and false positives as given by Eq. 3:

$$PRE = \frac{TP}{TP + FP} \tag{3}$$

Recall (REC) is the proportion of the true positives against the true positives and false negatives as given by Eq. 4:

$$REC = \frac{TP}{TP + FN} \tag{4}$$

F-measure takes values between 0 and 1. It is the harmonic mean of precision and recall as determined by Eq. 5:

$$F - measure = \frac{2 * PRE * REC}{PRE + REC} \tag{5}$$

4.3 Experimental Results

In Table 2, the predictive performances of three word embedding based feature representation schemes, namely, word2vec, global vectors (GloVe) and fastText have been presented in terms of F-measure values. For the results presented in Table 2, convolutional neural network has been utilized. For word2vec and fastText methods, continuous skip-gram and continuous bag of words (CBOW) schemes have been taken into consideration with varying vector sizes and dimensions of projection layers. In addition, different subsets of Twitter messages, ranging from 5000 to 50.000 are considered. In Table 2, subset1 represents subset of corpus with 5000 tweets, subset2 represents subset of corpus with 10000 tweets, subset3 represents subset of corpus with 15000 tweets, subset4 represents subset of corpus with 25000 tweets and finally, subset5 represents the entire corpus with 50.000 tweets.

As it can be observed from the results presented in Table 2, fastText based word embedding generally yield better predictive performance compared to global vectors (GloVe) and word2vec for text sentiment classification. The second highest predictive performance is generally obtained by GloVe based word embedding and the lowest predictive performance is generally obtained by word2vec based word embedding. Regarding the F-measure values in terms of different representation schemes, continuous bag of words generally yield better F-measure values compared to continuous skip-gram model. Regarding the vector size parameter of the empirical analysis, vector size of 300.0 yields better F-measure values, compared to vector size of 200.0. Regarding the dimension of vectors in the empirical analysis, the highest predictive performance is obtained for vector dimension of 300.0. The main findings of the empirical analysis have been summarized in Fig. 1.

Table 2. F-measure values obtained by different word-embedding schemes

Word Embedding	Scheme	Vector size	Dimension of projection layer	Subset#1	Subset#2	Subset#3	Subset#4	Subset#5
word2vec	Skip-gram	200.000	100.000	0.646	0.649	0.649	0.652	0.701
word2vec	Skip-gram	200.000	200.000	0.661	0.665	0.666	0.667	0.717
word2vec	Skip-gram	300.000	100.000	0.673	0.673	0.673	0.677	0.728
word2vec	Skip-gram	300.000	300.000	0.680	0.684	0.685	0.685	0.732
word2vec	CBOW	200.000	100.000	0.692	0.700	0.701	0.703	0.751
word2vec	CBOW	200.000	200.000	0.706	0.707	0.708	0.708	0.765
word2vec	CBOW	300.000	100.000	0.715	0.720	0.722	0.723	0.767
word2vec	CBOW	300.000	300.000	0.725	0.726	0.727	0.727	0.780
fastText	Skip-gram	200.000	100.000	0.837	0.837	0.840	0.841	0.916
fastText	Skip-gram	200.000	200.000	0.816	0.817	0.818	0.820	0.897
fastText	Skip-gram	300.000	100.000	0.806	0.808	0.808	0.811	0.874
fastText	Skip-gram	300.000	300.000	0.802	0.803	0.803	0.803	0.867
fastText	CBOW	200.000	100.000	0.795	0.798	0.799	0.800	0.863
fastText	CBOW	200.000	200.000	0.778	0.779	0.788	0.789	0.861
fastText	CBOW	300.000	100.000	0.773	0.774	0.775	0.777	0.860
fastText	CBOW	300.000	300.000	0.765	0.769	0.770	0.771	0.832
GloVe	–	200.000	100.000	0.727	0.731	0.734	0.735	0.784
GloVe	–	200.000	200.000	0.736	0.738	0.739	0.739	0.802
GloVe	–	300.000	100.000	0.741	0.746	0.749	0.753	0.806
GloVe	–	300.000	300.000	0.761	0.762	0.762	0.762	0.818

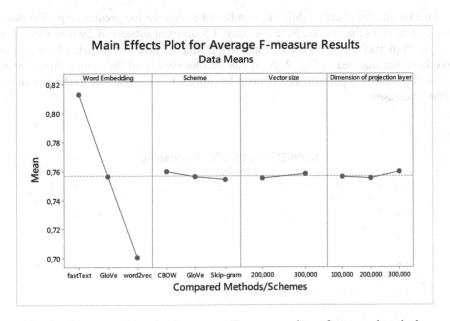

Fig. 1. The main effects plot for average F-measure values of compared methods.

In addition to deep learning based schemes presented in Table 2, the empirical analysis has also been given regarding the predictive performance of conventional supervised learning algorithms in conjunction with different word embedding schemes. In Table 3, the empirical results obtained by two supervised learning algorithms (namely, Naïve Bayes and support vector machines) with subset5 have been presented.

As it can be observed from the predictive performance results presented in Table 3, deep-learning based schemes (namely, convolutional neural network) yield better predictive performance compared to conventional supervised learning algorithms. Among the conventional learning algorithms compared in the analysis, support vector machines generally yield better predictive performance.

Table 3. F-measure values by supervised learning algorithms

Word embedding	Scheme	CNN	Support vector machines	Naive Bayes
word2vec	Skip-gram	0,701	0,615	0,615
word2vec	Skip-gram	0,717	0,625	0,619
word2vec	Skip-gram	0,728	0,648	0,623
word2vec	Skip-gram	0,732	0,675	0,642
word2vec	CBOW	0,751	0,678	0,657
word2vec	CBOW	0,765	0,690	0,681
word2vec	CBOW	0,767	0,749	0,738
word2vec	CBOW	0,780	0,763	0,752

To examine the effect of different number of tweets for the predictive performance of deep learning schemes, we have considered 5 different subsets of Twitter messages, ranging from 5000 to 50.000. The results regarding the different subsets of the corpus have been summarized in Fig. 2. As it can be observed from the results presented in Fig. 2, the predictive performance of deep-learning schemes enhances as the size of the corpus increases.

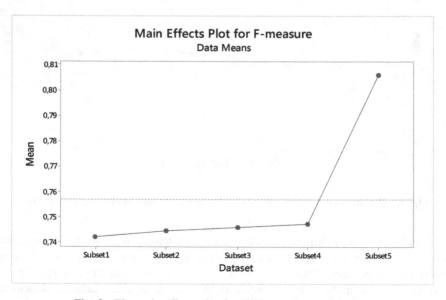

Fig. 2. The main effects plot for different subsets of dataset.

5   Conclusion

Sentiment analysis is an important research direction in natural language processing. The identification of sentiment can be useful for individual decision makers, business organizations and governments. Deep learning is a promising research field of machine learning. To build deep learning-based schemes for natural language tasks, word embeddings based representations play an essential role. In this regard, three well-known deep learning-based word embedding schemes, namely, word2vec, global vectors (GloVe) and fastText have been taken into consideration. In the empirical analysis, different subsets of Twitter messages, ranging from 5000 to 50.000 are taken into consideration. The prediction results obtained by deep-learning based schemes have been compared to conventional classifiers (such as, Naïve Bayes and support vector machines). The empirical analysis indicates that deep learning based schemes outperform conventional classification schemes (such as Naïve Bayes and support vector machines) for sentiment analysis. fastText based word embedding generally yield better predictive performance compared to global vectors (GloVe) and word2vec for text sentiment classification. In addition, the predictive performance of deep-learning schemes enhances as the size of the corpus increases.

References

1. Turney, P.D.: Thumbs up or thumbs down?: semantic orientation applied to unsupervised classification of reviews. In: Proceedings of the 40th Annual Meeting on Association for Computational Linguistics, pp. 417–424. Association for Computational Linguistics, July 2002
2. Somasundaran, S., Wiebe, J. Recognizing stances in ideological on-line debates. In: Proceedings of the NAACL HLT 2010 Workshop on Computational Approaches to Analysis and Generation of Emotion in Text, pp. 116–124. Association for Computational Linguistics, June 2010
3. Jo, Y., Oh, A.H.: Aspect and sentiment unification model for online review analysis. In: Proceedings of the Fourth ACM International Conference on Web Search and Data Mining, pp. 815–824. ACM, February 2011
4. Onan, A., Korukoğlu, S., Bulut, H.: A hybrid ensemble pruning approach based on consensus clustering and multi-objective evolutionary algorithm for sentiment classification. Inf. Process. Manage. **53**(4), 814–833 (2017)
5. Fersini, E., Messina, E., Pozzi, F.A.: Sentiment analysis: Bayesian ensemble learning. Decis. Support Syst. **68**, 26–38 (2014)
6. Liu, B.: Sentiment analysis and opinion mining. Synth. Lect. Hum. Lang. Technol. **5**, 1–167 (2012)
7. Onan, A., Korukoğlu, S.: A feature selection model based on genetic rank aggregation for text sentiment classification. J. Inform. Sci. **43**(1), 25–38 (2017)
8. Medhat, W., Hassan, A., Korashy, H.: Sentiment analysis algorithms and applications: a survey. Ain Shams Eng. J. **5**(4), 1093–1113 (2014)
9. Hailong, Z., Wenyan, G., Bo, J.: Machine learning and lexicon based methods for sentiment classification: a survey. In: 2014 11th Web Information System and Application Conference, pp. 262–265. IEEE, September 2014

10. Onan, A., Korukoğlu, S., Bulut, H.: A multiobjective weighted voting ensemble classifier based on differential evolution algorithm for text sentiment classification. Expert Syst. Appl. **62**, 1–16 (2016)
11. Deng, L., Yu, D.: Deep learning: methods and applications. Found. Trends Signal Process. **7** (3–4), 197–387 (2014)
12. Glorot, X., Bordes, A., Bengio, Y.: Domain adaptation for large-scale sentiment classification: a deep learning approach. In: Proceedings of the 28th International Conference on Machine Learning (ICML-11), pp. 513–520 (2011)
13. dos Santos, C., Gatti, M. Deep convolutional neural networks for sentiment analysis of short texts. In: Proceedings of COLING 2014, the 25th International Conference on Computational Linguistics: Technical Papers, pp. 69–78 (2014)
14. Tang, D., Wei, F., Qin, B., Liu, T., Zhou, M.: Coooolll: a deep learning system for twitter sentiment classification. In: Proceedings of the 8th International Workshop on Semantic Evaluation (SemEval 2014), pp. 208–212 (2014)
15. Severyn, A., Moschitti, A.: Twitter sentiment analysis with deep convolutional neural networks. In: Proceedings of the 38th International ACM SIGIR Conference on Research and Development in Information Retrieval, pp. 959–962. ACM, August 2015
16. Hu, Z., Hu, J., Ding, W., Zheng, X.: Review sentiment analysis based on deep learning. In: 2015 IEEE 12th International Conference on e-Business Engineering (ICEBE), pp. 87–94. IEEE, October 2015
17. Johnson, R., Zhang, T.: Effective use of word order for text categorization with convolutional neural networks. arXiv preprint arXiv:1412.1058 (2014)
18. Tang, D., Qin, B., Liu, T. Document modeling with gated recurrent neural network for sentiment classification. In: Proceedings of the 2015 Conference on Empirical Methods in Natural Language Processing, pp. 1422–1432 (2015)
19. Tang, D., Qin, B., Liu, T.: Learning semantic representations of users and products for document level sentiment classification. In: Proceedings of the 53rd Annual Meeting of the Association for Computational Linguistics and the 7th International Joint Conference on Natural Language Processing (Volume 1: Long Papers), vol. 1, pp. 1014–1023 (2015)
20. Ruder, S., Ghaffari, P., Breslin, J.G.: Insight-1 at semeval-2016 task 5: deep learning for multilingual aspect-based sentiment analysis. arXiv preprint arXiv:1609.02748 (2016)
21. Sun, X., Li, C., Ren, F.: Sentiment analysis for Chinese microblog based on deep neural networks with convolutional extension features. Neurocomputing **210**, 227–236 (2016)
22. Araque, O., Corcuera-Platas, I., Sanchez-Rada, J.F., Iglesias, C.A.: Enhancing deep learning sentiment analysis with ensemble techniques in social applications. Expert Syst. Appl. **77**, 236–246 (2017)
23. Paredes-Valverde, M.A., Colomo-Palacios, R., Salas-Zárate, M.D.P., Valencia-García, R.: Sentiment analysis in Spanish for improvement of products and services: A deep learning approach. Sci. Program. **2017**, 6 (2017)
24. Koppel, M.: Automatically categorizing written texts by author gender. Lit. Ling. Comput. **17**, 401–412 (2002)
25. Onan, A.: Sarcasm identification on twitter: a machine learning approach. In: Silhavy, R., Senkerik, R., Kominkova Oplatkova, Z., Prokopova, Z., Silhavy, P. (eds.) CSOC 2017. AISC, vol. 573, pp. 374–383. Springer, Cham (2017). https://doi.org/10.1007/978-3-319-57261-1_37
26. Rezaeinia, S.M., Ghodsi, A., Rahmani, R.: Improving the accuracy of pre-trained word embeddings for sentiment analysis. arXiv preprint arXiv:1711.08609 (2017)
27. Mikolov, T., Chen, K., Corrado, G., Dean, J.: Efficient estimation of word representations in vector space. arXiv preprint arXiv:1301.3781 (2013)

28. Zhang, L., Wang, S., Liu, B.: Deep learning for sentiment analysis: a survey. Wiley Interdiscip. Rev. Data Min. Knowl. Discov. **8**, e1253 (2018)
29. Bairong, Z., Wenbo, W., Zhiyu, L., Chonghui, Z., Shinozaki, T.: Comparative analysis of word embedding methods for DSTC6 end-to-end conversation modeling track. In: Proceedings of the 6th Dialog System Technology Challenges (DSTC6) Workshop (2017)
30. Bojanowski, P., Grave, E., Joulin, A., Mikolov, T.: Enriching word vectors with subword information. arXiv preprint arXiv:1607.04606 (2016)
31. Pennington, J., Socher, R., Manning, C.: Glove: global vectors for word representation. In: Proceedings of the 2014 Conference on Empirical Methods in Natural Language Processing (EMNLP), pp. 1532–1543 (2014)
32. Young, T., Hazarika, D., Poria, S., Cambria, E.: Recent trends in deep learning based natural language processing. IEEE Comput. Intell. Mag. **13**(3), 55–75 (2018)
33. Kilimci, Z.H., Akyokus, S.: Deep learning-and word embedding-based heterogeneous classifier ensembles for text classification. Complexity **2018**, 10 (2018)
34. Cireşan, D., Meier, U., Schmidhuber, J.: Multi-column deep neural networks for image classification. arXiv preprint arXiv:1202.2745 (2012)

A Cluster-Based Machine Learning Model for Large Healthcare Data Analysis

Fatemeh Sharifi[1(✉)], Emad Mohammed[2], Trafford Crump[3], and Behrouz H. Far[1]

[1] Department of Electrical and Computer Engineering, University of Calgary, Calgary, AB, Canada
{fatemeh.sharifi1,far}@ucalgary.ca
[2] Faculty of Engineering, Lakehead University, Thunder Bay, ON, Canada
emohamme@lakeheadu.ca
[3] Department of Surgery, University of Calgary, Calgary, AB, Canada
tcrump@ucalgary.ca

Abstract. There is huge growth in the amount of patient survey data being generated in healthcare industries and hospitals. Curse of dimensionality is a barrier to extracting useful information from patient survey data which can help in the treatment and care of patients. It is paramount to have methods to find importance of features based on such huge volumes of stored information for the desired outputs. The health-related quality of life (HRQOL) is a powerful paradigm to help reaching such a desired output, measuring as patient satisfaction. In such scenarios, it is difficult to investigate the features, out of such high-dimensional data, that could best represent desired output and explain them so that such features can be used in the future at the point f care. In this paper we propose a Cluster-based Random Forest (CB-RF) method to particularly exploit the most important features for the desired output, which is Expanded Prostate Index Composite-26 (EPIC-26) domain scores. EPIC-26 is being used for assessing a range of HRQOL issues related to the diagnosis and treatment of prostate cancer. Different feature extraction methods are applied to extract features and the best method is the proposed CB-RF model which could find the most important features (10 or less) out of over 1500 features that can be used to accurately estimate patient with their EPIC-26 values with on average 85% coefficient of correlation between predicted and observed values of real dataset including 5093 patients.

Keywords: Machine learning · Big data · Patient quality of life · Dimension reduction

1 Introduction

Tremendous progress in data collection devices enables hosting vast data modalities in health care. Electronic Health Records (EHRs) are commonly used in the diagnosis and management of patients' diseases. The Office of National Coordinator reported that EHRs have been deployed in more than 95% of hospitals

© Springer Nature Switzerland AG 2019
M. Younas et al. (Eds.): Innovate-Data 2019, CCIS 1054, pp. 92–106, 2019.
https://doi.org/10.1007/978-3-030-27355-2_7

in the United States in 2016 [17]. The interpretation of structured information from EHRs can guide clinical care.

Patient-reported outcomes (PROs) is one example of data that is potentially useful for providing care for patients. PROs can be used to measure HRQOL and monitor symptom severity, which can be used to improve patient care [2,12] and facilitate patient-provider communication [23].

The increasing availability of data in health care has made the need for developing machine learning methods to extract the information and features from the big data a necessity. However, it is usually very difficult to extract useful information from EHRs because the nature of EHR data is highly dimensional. High-dimensional medical datasets present many mathematical challenges and have given impressive rise to new theoretical developments [4]. One of the problems with high-dimensional datasets is that, specifically in medical cases, not all the measured variables are **important** for understanding and predicting the phenomena of interest. Although various recent works have been proposed novel methods with high accuracy from high-dimensional data [3], in health care, it is still of interest to reduce the dimension of the original data to only those that have more contribution in case of the desired output prior to any modeling of the data. The task of dimension reduction is necessary to find specific variables out of all records related to a person considering a specific desired output. In these cases, however, the dimension reduction can be very complex due to the large number of dimensions, incomplete data, and multiple ways to represent the same information.

In this paper, we propose a parallel processing technical approach in which the unsupervised and supervised learning are combined to decrease the overall error for predicting HRQOL using diagnostic, clinical, and PRO data by reducing the number of dimensions to the most important features from big health care data. There are several challenges to achieving this aim.

1.1 Technical Significance

To the best of our knowledge, our work is the first that proposes to do supervised learning on top of unsupervised learning in high-dimensional semi-structured patient survey data to extract the most important features for EPIC-26 Values which can predict it with high coefficient of correlation between predicted and observed values. This paper has three main contributions. In the first place, extracting minimal features with the highest possible prediction power out of more than 1500 features for each domain in EPIC-26. Secondly, estimating the actual value of the EPIC-26 on the patients (i.e. a continuous number between 0 and 100) not only the classification (i.e. Yes or No for patient satisfaction). And finally, Conducting a large-scale empirical study on a real-word health care dataset with 5093 patients. The details of all technical methods is provided in Sect. 4.

1.2   Clinical Relevance

According to the Canadian Cancer Society, prostate cancer (PCa) is the most common cancer diagnosis in men - 1 in 8 men will be diagnosed with the disease in their lifetime, or approximately 21,500 per year [5]. These men face a decision about their treatment. Surgery, radiation, chemotherapies, and active surveillance are all options. All of these options are associated with risks for potential benefits and harms and can potentially have dramatic consequences on a man's quality-of-life [13,14,16] and result in an increased use in healthcare services [21].

While predictive models for PCa exist (e.g., the nomograms developed by Memorial Sloan Kettering Cancer Centre [15]), they are limited in several important ways. The first is in their generalizability to the Canadian context, being based on clinical data from other health care jurisdictions that may approach PCa diagnosis and treatment differently than in Canada. The second is that these predictive models do not include PROs or health care utilization outcomes.

2   Data

This study is based on a secondary analysis of data prospectively collected from participants enrolled in the Alberta Prostate Cancer Research Initiative (APCaRI) study [1] which includes 5093 patients with 1592 dependent and independent variables. The APCaRI study is enrolling the population of men undergoing diagnosis for prostate cancer in Calgary and Edmonton, Canada. These patients were referred for a diagnostic biopsy based on conventional clinical guidelines (e.g., elevated PSA and/or abnormal digital rectal examination). To be eligible for the APCaRI study, men must be over 18 years of age, speak English or have a translator available, and not have had a prior prostate cancer diagnosis. Participants in the APCaRI study consent to having their data used for secondary studies. Only participants whom have provided this consent are included in the analytic data set used for this study. To be eligible for this study, participants had to have been diagnosed with prostate cancer and completed the EPIC-26. This study is approved by the Conjoint Health Research Ethics Board at the University of Calgary.

2.1   Data Collection

Participants' demographic characteristics were collected during the in-person interview prior to biopsy. Trained data collectors abstracted participants [5]; clinical data from laboratory, pathology, or treatment reports. All data were entered into a REDCap database [11]. All personal identifiers were removed from the APCaRI study database to our research team for analysis.

3   Desired Output, EPIC-26

The EPIC-26 questionnaire is a valid and reliable subjective measure HRQOL in men undergoing prostate cancer treatment [22]. The items of EPIC-26 are

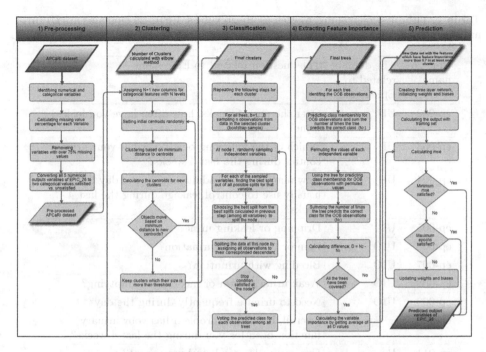

Fig. 1. Flowchart of the approach described in this paper. After collecting the APCaRI data, the steps are (1) Pre-processing (2) Clustering (3) Classification (4) Extracting feature importance and (5) Prediction.

responded to using either a four- or five-item Likert scale. Responses are transformed to a 0–100 scale, with higher scores representing less symptom severity. Using the available scoring instructions [20], items are grouped into one of five domains: urinary incontinence (UI) (4 items), urinary irritative/obstructive (UIO) (4 items), bowel (B) (6 items), sexual (S) (6 items), and hormonal (H) (5 items) [25]. As shown in Table 1 there are 26 items in which one item (i.e., "Overall, how big a problem has your urinary function been for you during the last 4 weeks?") is not included in any domain. For each domain, item scores are averaged to calculate the domain summary score. Participants completed the EPIC-26 either in-person (if the participant is willing to come in to the clinic), over a secure website, or by filling the questionnaire mailed to them.

4 Method

Before we start to explain proposed method, it is necessary to understand what characteristics are desired for the model. Epic-26 values for each domain for all the patients are a continuous value between 0 and 100, with 0 indicating the least satisfaction and 100 shows the highest. All of these 26 features which are the influential features for Epic-26 domains values are hidden in the 1592

Table 1. The questions related to each EPIC-26 values. EPIC-26 Domains are UI (urinary incontinence), UIO (Urinary Irritative/Obstructive), B (Bowel) and S (sexual).

Epic-26 name	The domain related to Epic name	The question related to Epic name
ep-002	UI	Over the past 4 weeks, how often have you leaked urine?
ep-003	UI	Which of the following best describes your urinary control during the last 4 weeks?
ep-004	UI	How many pads or adult diapers per day did you usually use to control leakage during the last 4 weeks?
ep-005	UI	Dripping or leaking urine?
ep-006	UIO	Pain or burning on urination?
ep-007	UIO	Bleeding with urination?
ep-008	UIO	Weak urine stream or incomplete emptying?
ep-009	UIO	Need to urinate frequently during the day?
ep-010	_	Overall, how big a problem has your urinary function been for you during the last 4 weeks?
ep-011	B	Urgency to have a bowel movement?
ep-012	B	Increased frequency of bowel movements?
ep-013	B	Losing control of your stools?
ep-014	B	Bloody stools?
ep-015	B	Abdominal/Pelvic/Rectal pain?
ep-016	B	Overall, how big a problem have your bowel habits been for you during the last 4 weeks?
ep-017	S	Your ability to have an erection?
ep-018	S	Your ability to reach orgasm (climax)?
ep-019	S	How would you describe the usual QUALITY of your erections during the last 4 weeks?
ep-020	S	How would you describe the FREQUENCY of your erections during the last 4 weeks?
ep-021	S	Overall, how would you rate your ability to function sexually during the last 4 weeks?
ep-022	S	Overall, how big a problem has your sexual function or lack of sexual function been for you during the last 4 weeks?
ep-023	H	Hot flashes?
ep-024	H	Breast tenderness/enlargement?
ep-025	H	Feeling depressed?
ep-026	H	Lack of energy?
ep-027	H	Change in body weight?

features we have in the dataset. By taking this into account, in this section, we explain our cluster-based random forest approach for health related quality of life prediction, as per Fig. 1, by first clustering the patients and then using random forest method for feature engineering and finally applying the prediction technique on the patients considering only the extracted features to evaluate the feature extraction method.

4.1   Patient Clustering

The purpose of performing clustering before classification can be explained as follows. The patient survey data set is an almost highly imbalanced data. Table 2 demonstrate that there are 2.62%, 1.19%, 0.69%, 26.97% and 1.2% dissatisfied responses (i.e. less than 50) for UI, UIO, B, S and H, respectively. Therefore, to prevent classification model to overlook the unsatisfied responses which are minority, we utilized K-means which is a popular clustering methods. Using the clustering, patients which are similar (i.e. unsatisfied patients) would probably end up in the same cluster. Then considering the unsatisfied clusters separately for classification, will help to find the influential features on minority responses as well as on majority responses in which include the satisfied patients (i.e. there EPIC-26 values are more than 50). In this study we assigned N new columns for categorical features with N levels so can be used in k-means. In the K-means clustering, first, k centroids are randomly selected. The number of k, the cluster numbers, is decided with the elbow method. Each point which is patient in our study, decides its centroid nearest to it according to the Euclidean distance. After each of the patients in the dataset is assigned to one of the k clusters, the centroid of each cluster is updated. This procedure is repeated until the centroids are not changed anymore. At the end of this part the clusters are formed.

Table 2. Statistical characteristics of patients for all the EPIC-26 domains score quartiles. EPIC-26 Domains are UI (urinary incontinence), UIO (Urinary Irritative/Obstructive), B (Bowel) and S (sexual).

Range of score	No. patients for each EPIC-26 domain				
	UI (%)	UIO (%)	B (%)	S (%)	H (%)
[0–25)	17 (0.78)	2 (0.09)	0 (0)	279 (12.82)	5 (0.23)
[25–50)	40 (1.84)	24 (1.1)	15 (0.69)	308 (14.15)	21 (0.97)
[50–75)	179 (8.23)	251 (11.53)	69 (3.17)	356 (16.36)	128 (5.88)
[75–100]	1372 (63.05)	1630 (74.91)	1879 (86.35)	872 (40.07)	1076 (49.45)
Missing values	568 (26.1)	269 (12.36)	213 (9.79)	361 (16.59)	946 (43.47)

Elbow Method: Elbow method [18] is a visual method to determine approximately the best number of clusters. Reason being is that starting with $K = 2$, and increasing by 1 in each step, we want to select a K where the cost, which is within cluster sum of squares (WSSE) in this paper, before that drops significantly but beyond the "elbow" it reaches a plateau when you increase it further. In the Fig. 2 we can see in our cases using 18 clusters should be the right number of clusters for Epic-26 variables, as it is the elbow of the curve.

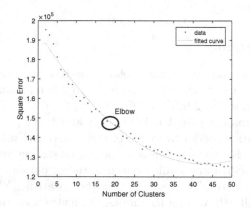

Fig. 2. Elbow method for finding the best number of clusters.

4.2 Patient Classification

At the end of previous step we ended up having patients in 18 different clusters. To enhance the classification output based on the satisfaction of the patients, we assigned value "Yes", indicating their satisfaction, to the Epic-26 variables if their value is grater than or equal to 50 and "NO" if their value is less than 50. The new features will be used as outcome of interest only for training our classifier not for using it in the prediction part. In other words, the prediction algorithm will estimate the actual values of EPIC-26 (i.e. the continuous number between 0 and 100) not the classified values (i.e. Yes or No). The selected classifier in this step is random forest as it is a reasonable suggestion for our semistructured high-dimensional dataset which most of its variables are categorical (i.e. radio, checkbox, yesno, etc.). Random Forest is a tree-based algorithm that combines the results of many weak trees/learners to have improvement in predication accuracy.

The RF mainly requires the definition of two parameters to generate importance of features, the number of classification trees desired (ntree) and the number of columns to randomly select at each level (mtry). In this study, several RF models were constructed using the selected features, each with ntree of 1000 and mtry equaling to the square root of the number of input variables [6]. Furthermore, RF implementation on H2O [10] was used because of its flexibility and processing speed.

4.3   Extracting Important Features

Random Forest model provides an expert way to measure features' importance. The importance for each variable, is calculated by permutation. It describes the difference in prediction accuracy before and after permuting the variable of interest, averaged over all trees [3], by representing the mean decrease in classification accuracy after permuting each input variable over all trees. To determine the permutation importance of a variable, it should be permuted while keeping all other variables in the out-of-bag (OOB) samples unchanged, then running all trees on the OOB samples and finally calculating how much the squared error improved as a result. The average of this number over all trees in the forest is the importance score for the variable. In our study, the top important features which are considered those with importance more than 0.7 (it is selected by try and error to get the most efficient threshold which the extracted features does not exceed 10 features) in at least one of the clusters.

4.4   Prediction of the EPIC-26 Actual Values

Artificial Neural Networks (NNs) have been applied to the extracted features from the previous step to predict the actual number of EPIC-26 for each domain. The data were divided to 70%, 15% and 15% for training, validation and testing, respectively. We repeated the above training and testing process 10 times and the results are reported in Table 3.

Neural Network Set up: The NN we employed in this study is a simple multi-layer perceptron with 3 layers containing 10 neurons in hidden layer. The network has 10 inputs which are the minimal set of important features extracted by each of the methods, and one output is a continuous number for EPIC-26 domain value. We used the logistic function as the activation function inside the neurons with backpropagation as the learning mechanism.

5   Results

5.1   Baseline

Among different methods for feature extraction and dimension reduction, we study the use of: (i) Random Forest on the dataset without clustering; (ii) PCA on the dataset without clustering; (iii) Deep learning on the clustered data. For the baseline, a 3 layer NN is used as the prediction method (as a powerful paradigm for prediction) for all the feature extraction/dimension reduction methods.

5.2   Evaluation Approach/Study Design

In this section, we explain our evaluation of the proposed cluster-based feature extraction technique. We used NN for predicting EPIC-26 exact values to evaluate if the extracted features are predictive or not.

Objective. The objective of this experiment is to propose a parallel processing technical approach in which the unsupervised learning and supervised learning are combined to decrease the overall error for predicting HRQOL using diagnostic, clinical, and PRO data by reducing the number of dimensions to identify the most important features from the dataset.

Evaluation Metric. As we are predicting a continuous value for each domains of EPIC-26, the metric used in this paper for assessing the effectiveness of the predictive model is Coefficient of correlation (R) between predicted and observed values in EPIC-26 domains. In case of using R-squared, it can be calculated as $R * R$ or R^2. Correlation coefficient reflects the level of direct relationship between two components, i.e., how much the variables are associated, and the values are in the range of +1 to −1. An association of +1 suggests that there is an extraordinary positive direct relationship between elements. In other words, the predicted values is the same as the actual value, while a negative association demonstrates disagreement between two values.

Research Questions

RQ1: Can a basic Random Forest algorithm with a simple NN prediction method improve the effectiveness of a traditional dimension reduction, PCA, with the same prediction method? In this research question, we compare the prediction results of Random Forest as part of our method with the PCA method.

RQ2: Does Clustering the patients first and then extracting the features, improve the prediction coefficient of correlation? In this research question, we upgrade the simple RF method, which applies the extracting feature importance on the whole data, to a cluster-based Random Forest (CB-RF). The proposed method first starts with clustering the patients and then applies the feature extraction models on the patients in the clusters separately. In this research question we compare CB-RF with original random forest (ORF) on the whole data not the clustered one.

RQ3: Keeping the clustering part of the method, are the features extracted with Random Forest, better than features extracted with another powerful feature extraction method, deep learning which has been applied on the clustered data as well? In this research question, we investigate the EPIC-26 prediction ability of Random Forest vs. Deep learning; However, both the algorithms are applied on the clustered patients for the sake of consistency.

Deep Learning Set Up: Deep learning aims at learning feature hierarchies with features from higher levels formed by the composition of lower level features. In this paper we use a deep neural network (DNN) as a conventional multilayer perceptron (MLP)) [19]. They include learning methods for a wide array of deep architectures, including neural networks with many hidden layers [24] [9].

Table 3. Coefficient of correlation between predicted and observed values in EPIC-26 domains. Neural Network (NN) is used for predicting EPIC-26 domains using important features calculated with different features extraction techniques: PCA, ORF (Original Random Forest), CB-DL (Cluster-based Deep Learning), CB-RF (Cluster-Based Random Forest) - EPIC-26 Domains are UI (urinary incontinence), UIO (Urinary Irritative/Obstructive), B (Bowel) and S (sexual). For NN, R values of 10 runs and their average (AVG) and Standard Deviation (SD) are reported.

Method	EPIC-26 Domain	Run 1	Run 2	Run 3	Run 4	Run 5	Run 6	Run 7	Run 8	Run 9	Run 10	AVG	SD
PCA	UI	−0.0944	−0.1095	0.1393	−0.1256	−0.0627	−0.1049	0.0096	−0.1703	0.0466	0.0701	−0.0402	0.1006
	UIO	0.2387	0.1768	0.0585	0.1067	0.1667	0.3139	0.1754	0.2796	−0.0041	0.0813	0.1594	0.1006
	B	−0.0122	−0.0073	−0.1144	−0.1075	0.1474	0.2904	0.1535	0.0166	0.1223	0.1225	0.0611	0.1279
	S	0.3601	0.088	0.1011	0.2488	0.1474	−0.1366	0.4373	0.149	0.235	0.4843	0.2114	0.1844
	H	0.1065	−0.3202	0.053	0.3192	0.2827	−0.1631	0.0604	0.2468	0.0188	0.0437	0.0648	0.1972
ORF	UI	0.7204	0.7081	0.7032	0.5946	0.7403	0.6875	0.7871	−0.6036	0.6008	0.6299	0.5568	0.4123
	UIO	0.8536	0.8545	0.879	0.8719	0.874	0.8786	0.8608	0.873	0.8796	0.882	0.8707	0.0106
	B	0.8538	0.588	0.8782	0.8024	0.8558	0.8879	0.8878	0.872	0.9093	0.8143	0.8350	0.0929
	S	0.9431	0.9354	0.936	0.9351	0.9276	0.9479	0.9365	0.9479	0.9534	0.9366	0.9400	0.0078
	H	0.403	0.5316	0.4414	0.5906	0.6003	0.5648	0.6446	0.6824	0.6163	0.5433	0.5618	0.0867
CB-DL	UI	0.9258	0.9645	0.9329	0.965	0.9548	0.9069	0.956	0.9229	0.901	0.9014	0.9331	0.0256
	UIO	0.5046	0.472	0.4939	0.4142	0.5399	0.4604	0.5321	0.6296	0.3404	0.5767	0.4964	0.0819
	B	0.1858	0.193	0.189	0.1504	0.2134	0.269	0.0512	0.0895	0.2009	0.1514	0.1694	0.0624
	S	0.9303	0.9474	0.9441	0.9562	0.9455	0.9463	0.9438	0.9415	0.9577	0.935	0.9448	0.0083
	H	0.1119	0.105	0.0525	−0.1691	0.1855	0.1354	0.1086	−0.0785	−0.0203	0.1021	0.0533	0.1096
CB-RF	UI	0.993	0.9878	0.9917	0.9873	0.986	0.9936	0.9175	0.9915	0.988	0.9921	0.9829	0.0231
	UIO	0.7724	0.8073	0.7585	0.789	0.7866	0.7201	0.6928	0.43441	0.7938	0.7715	0.7451	0.0741
	B	0.8679	0.8622	0.8698	0.2506	0.928	0.6722	0.9212	0.8841	0.895	0.8722	0.8023	0.2065
	S	0.88207	0.8543	0.8786	0.9061	0.895	0.8741	0.9069	0.8411	0.8325	0.8488	0.8719	0.0267
	H	0.8711	0.8896	0.8182	0.8886	0.895	0.9135	0.4976	0.8946	0.9102	0.9163	0.8495	0.1268

For Deep Learning, variable importance is calculated using the Gedeon [8] method. It is a modification of Garson method [7] in which the hidden layer weights are partitioned into components associated with each input node. After that the percentage of all hidden nodes weights assigned to a particular input node is used to measure the importance of that input variable. Mainly it is based on using the weight matrix of the trained neural network itself to determine which inputs are significant.

In RQ3, we define a standard deep learning with multilayer perceptron with 5 hidden layers, each containing 100 hidden nodes. The activation function used in the layers are $Tanh$. We use stochastic gradient descent using backpropagation as training algorithm. For improving generalization we specify the input layer dropout ratio as 0.2.

5.3   Case Study Results

In this section, we explain and discuss the results of the experiment and answer the three research questions.

RQ1: To answer this research question, we compare the original RF used in our proposed method with the PCA as a standard statistical technique for dimension reduction. In both of the methods the whole dataset is used without any clustering, and then we used NN as the prediction model in both methods. The idea is to see the effect of RF without giving much advantages to the clustering. Looking at Fig. 3, our first observation is that overall the PCA's R values are very low. Moreover, the figure shows that the RF approach is quite effective in the predicting quality of life variables.

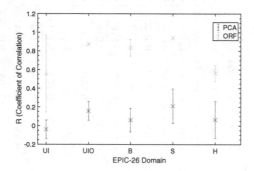

Fig. 3. Mean and standard deviations of coefficient of correlation for PCA vs a simple (Original) Random Forest (RF). EPIC-26 Domains are UI (urinary incontinence), UIO (Urinary Irritative/Obstructive), B (Bowel) and S (sexual).

The previous analysis show that the results of the Random Forest is better than PCA which is in favor of the RQ1. It also indicates that we can improve the basic PCA feature extraction technique using a simple RF with NN prediction model. This is quite interesting since it shows that PCA cannot work effectively on dataset which mostly is categorical and a simple RF brings extra knowledge about the features. This knowledge is basically the importance of the features which can not be seen in the PCA model. However, the RF that was used in RQ1 was quite simple. So the next question will be whether a more complex feature extraction can help in prediction or not, which is studied in RQ2.

RQ2: The objective of RQ2 is to study the cluster-based measure in more details. We want to know whether RF can be improved if we cluster all the patients and then extract feature importance in each cluster, or not. This basically tries to improve the RF without using new feature extraction models. Focusing on RQ2, given that in many outputs RF already provides high coefficient correlation, Fig. 4 shows that CB-RF can improve it effectively, meaning that the most significant information, in terms of the important features that

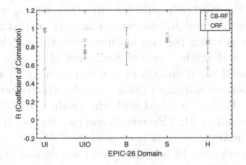

Fig. 4. Mean and standard deviations of coefficient of correlation for RF vs the proposed method, Cluster-Based Random Forest (CB-RF). EPIC-26 Domains are UI (urinary incontinence), UIO (Urinary Irritative/Obstructive), B (Bowel) and S (sexual).

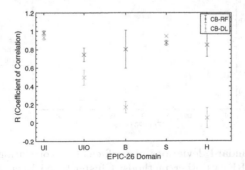

Fig. 5. Mean and standard deviations of coefficient of correlation for Cluster-based Deep Learning (CB-DL) vs the proposed method, Cluster-Based Random Forest (CBRF). EPIC-26 Domains are UI (urinary incontinence), UIO (Urinary Irritative/Obstructive), B (Bowel) and S (sexual).

Table 4. P-value of Cluster Based Random Forest (CB-RF) and other dimension reduction techniques.

Method	PCA	ORF	CB-DL	CB-RF
PCA	1	1.92E−25	8.37E−10	7.52E−28
ORF	1.92E−25	1	0.000173	0.018507
CB-DL	8.37E−10	0.000173	1	3.76E−08
CB-RF	7.52E−28	0.018507	3.76E−08	1

contribute to predicting the desired output, are common in patients in the same clusters, explaining why CB-RF is outperforming simple RF itself.

RQ3: In RQ3, we study the same question as RQ2, but in the context of different feature extraction models. In other words, we try to improve the CB-RF by a better feature extraction model. To answer RQ3, we replace RF with Deep Learning, retaining the cluster-based idea, in other words, we will cluster the patients and then use DL and RF to extract important features from clusters. Looking at Fig. 5, we see that the two feature extracting models, in the context of NN prediction, have different levels of coefficient of correlation. It shows that only being cluster-based is not enough even when we have an accepted and popular method which is deep learning in this case. Another observation is that the proposed method is the best method with the prediction R (correlation coefficient) of 85% (R-squared 72%). This conclusion can also be drawn by looking at Fig. 6 and Table 3, the improvement is on average 76%, 10% and 33% on PCA, RF and CB-DL, respectively. In addition, the low p-values reported in Table 4, all less than 0.05, show that the improvements are not random or by chance.

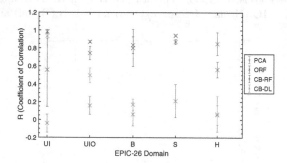

Fig. 6. Mean and standard deviations of coefficient of correlation for Cluster-based Random Forest (CB-RF) vs other methods: Cluster-based Deep Learning (CB-DL), PCA and Random Forest (RF). EPIC-26 Domains are UI (urinary incontinence), UIO (Urinary Irritative/Obstructive), B (Bowel) and S (sexual).

6   Conclusions and Related Works

Extracting important features is an expensive and tedious task for big data mining projects, which is essential for developing fast and high accuracy prediction models. In this paper we developed a CB-RF model to identify the importance of features from highly dimensional dataset. We then proposed a model to provide personalized service to patients by predicting the desired output for the EPIC-26 domains. Finally, we used three different feature extraction models to find the important features for the desired output, separately. We demonstrated the application of this approach using diagnostic, clinical, and PRO data extracted from patients diagnosed with prostate cancer. We used CB-RF to identify the most important features for predicting patients' HRQOL. Based on the empirical study results, we showed that a RF combined with patient clustering in a neural network model can accurately predict all variables of HRQOL with impressive improvement on the predicted results performance in comparison with three other models. In addition, the proposed model was completely on H2O, which has the advantage of both run time (less than two minutes) and code complexity for both clustering and feature extraction. To the best of our knowledge, this work is the first use of H2O on feature extraction and predicted models in medical domain and has shown promising results, which we are planning to replicate on different data sets and expand on different cluster-based feature extracting models to more accurately extract features from big datasets.

References

1. APCARI: Home-apcari. https://apcari.ca/
2. Basch, E., et al.: Adverse symptom event reporting by patients vs clinicians: relationships with clinical outcomes. J. Natl. Cancer Inst. **101**(23), 1624–1632 (2009)
3. Breiman, L.: Random forests. Mach. Learn. **45**(1), 5–32 (2001)
4. Breiman, L., Friedman, J., Stone, C.J., Olshen, R.A.: Classification and Regression Trees. Chapman and Hall, New York (1984)
5. Canadian Cancer Society: Prostate cancer statistics - Canadian Cancer Society. http://www.cancer.ca/en/cancer-information/cancer-type/prostate/statistics/?region=ab
6. Chan, J.C.W., Paelinckx, D.: Evaluation of random forest and adaboost tree-based ensemble classification and spectral band selection for ecotope mapping using airborne hyperspectral imagery. Remote Sens. Environ. **112**(6), 2999–3011 (2008)
7. Garson, G.D.: Interpreting neural-network connection weights. AI Expert **6**(4), 46–51 (1991)
8. Gedeon, T.D.: Data mining of inputs: analysing magnitude and functional measures. Int. J. Neural Syst. **8**(02), 209–218 (1997)
9. Glorot, X., Bengio, Y.: Understanding the difficulty of training deep feedforward neural networks. In: Proceedings of the Thirteenth International Conference on Artificial Intelligence and Statistics, pp. 249–256 (2010)
10. H2O.ai: Home - h2o.ai. https://www.h2o.ai/
11. Harris, P.A., Taylor, R., Thielke, R., Payne, J., Gonzalez, N., Conde, J.G.: Research electronic data capture (REDcap)–a metadata-driven methodology and workflow process for providing translational research informatics support. J. Biomed. Inform. **42**(2), 377–381 (2009)

12. Henry, J., Pylypchuk, Y., Searcy, T., Patel, V.: Adoption of electronic health record systems among us non-federal acute care hospitals: 2008–2015. ONC Data Brief **35**, 1–9 (2016)

13. Herschorn, S., Gajewski, J., Schulz, J., Corcos, J.: A population-based study of urinary symptoms and incontinence: the Canadian urinary bladder survey. BJU Int. **101**(1), 52–58 (2008)

14. Korfage, I.J., Essink-Bot, M.L., Janssens, A.C.J.W., Schröder, F.H., De Koning, H.J.: Anxiety and depression after prostate cancer diagnosis and treatment: 5-year follow-up. Br. J. Cancer **94**(8), 1093 (2006)

15. Memorial Sloan Kettering Cancer Center: Prostate cancer nomograms — memorial sloan kettering cancer center. https://www.mskcc.org/nomograms/prostate

16. Michaelson, M.D., Cotter, S.E., Gargollo, P.C., Zietman, A.L., Dahl, D.M., Smith, M.R.: Management of complications of prostate cancer treatment. CA: A Cancer J. Clin. **58**(4), 196–213 (2008)

17. Office of National Coordinator: Office of the national coordinator for health information technology (2016). https://dashboard.healthit.gov/quickstats/pages/FIG-Hospital-Progress-to-Meaningful-Use-by-size-practice-setting-area-type.php

18. Ng, A.: Clustering with the k-means algorithm. Machine Learning (2012)

19. Rosenblatt, F.: Principles of neurodynamics. Perceptrons and the theory of brain mechanisms. Cornell Aeronautical Lab Inc., Buffalo, NY (1961)

20. Sanda, M., Wei, J., Litwin, M.: Scoring instructions for the expanded prostate cancer index composite short form (EPIC-26). https://medicine.umich.edu/sites/default/files/content/downloads.Scoring%20Instructions%20for%20the%20EPIC%2026

21. Stokes, M.E., Black, L., Benedict, A., Roehrborn, C.G., Albertsen, P.: Long-term medical-care costs related to prostate cancer: estimates from linked seer-medicare data. Prostate Cancer Prostatic Dis. **13**(3), 278 (2010)

22. Szymanski, K.M., Wei, J.T., Dunn, R.L., Sanda, M.G.: Development and validation of an abbreviated version of the expanded prostate cancer index composite instrument for measuring health-related quality of life among prostate cancer survivors. Urology **76**(5), 1245–1250 (2010)

23. Velikova, G., et al.: Measuring quality of life in routine oncology practice improves communication and patient well-being: a randomized controlled trial. J. Clin. Oncol. **22**(4), 714–724 (2004)

24. Vincent, P., Larochelle, H., Bengio, Y., Manzagol, P.A.: Extracting and composing robust features with denoising autoencoders. In: Proceedings of the 25th International Conference on Machine Learning, pp. 1096–1103. ACM (2008)

25. Wei, J.T., Dunn, R.L., Litwin, M.S., Sandler, H.M., Sanda, M.G.: Development and validation of the expanded prostate cancer index composite (EPIC) for comprehensive assessment of health-related quality of life in men with prostate cancer. Urology **56**(6), 899–905 (2000)

Satire Detection in Turkish News Articles: A Machine Learning Approach

Mansur Alp Toçoğlu[1] and Aytuğ Onan[2(✉)]

[1] Faculty of Technology, Department of Software Engineering,
Manisa Celal Bayar University, 45400 Manisa, Turkey
mansur.tocoglu@cbu.edu.tr
[2] Faculty of Engineering and Architecture,
Department of Computer Engineering, İzmir Katip Çelebi University,
35620 İzmir, Turkey
aytug.onan@ikc.edu.tr

Abstract. With the advances in information and communication technologies, an immense amount of information has been shared on social media and microblogging platforms. Much of the online content contains elements of figurative language, such as, irony, sarcasm and satire. The automatic identification of figurative language can be viewed as a challenging task in natural language processing, where linguistic entities, such as, metaphor, analogy, ambiguity, irony, sarcasm, satire, and so on, have been utilized to express more complex meanings. The predictive performance of sentiment classification schemes may degrade if figurative language within the text has not been properly addressed. Satirical text is a way of figurative communication, where ideas/opinions regarding a people, event or issue is expressed in a humorous way to criticize that entity. Satirical news can be deceptive and harmful. In this paper, we present a machine learning based approach to satire detection in Turkish news articles. In the presented scheme, we utilized three kinds of features to model lexical information, namely, unigrams, bigrams and tri-grams. In addition, term-frequency, term-presence and TF-IDF based schemes have been taken into consideration. In the classification phase, Naïve Bayes, support vector machines, logistic regression and C4.5 algorithms have been examined.

Keywords: Satire identification · Fake news · Machine learning

1 Introduction

Satire can be defined as the use of humor, irony, exaggeration, or ridicule to criticize people, services, event or issue [1]. Satirical text is a form of nonliteral communication, where people express criticism regarding people or issues, particularly related to contemporary politics and other issues, in a humorous way. With the advances in information and communication technologies, large amount of user-generated information available on the Web. Sentiment analysis is the process of identification of subjective information towards an entity, event or issue, is a promising research field in natural language processing. Sentiment analysis on the user-generated information on the social media platforms can be utilized to identify sentiment orientation towards an

M. Younas et al. (Eds.): Innovate-Data 2019, CCIS 1054, pp. 107–117, 2019.
https://doi.org/10.1007/978-3-030-27355-2_8

entity, service or product, which may be an essential source of information for business organization, governments, and individual decision makers [2, 3]. Hence, the automatic identification of subjective information may be utilized to obtain structured knowledge, which can serve as an important source to build decision support systems.

Much of the online content available on the Web consists of figurative and nonliteral linguistic elements, such as metaphor, analogy, ambiguity, irony, sarcasm and satire. The automatic identification of figurative language can be viewed as a challenging task in natural language processing. In contrast to literal language, figurative language utilizes linguistic elements, including irony, sarcasm and satire to express more complex issues, that may be difficult to identify for not only computers, but also for human beings. In addition, the predictive performance of sentiment classification schemes may degrade if figurative language within the text has not been properly addressed [4].

The progress on Internet technology has sparked the growth of information sharing in forms of news, as well. Figurative language may be also encountered in news articles shared on the Web. News satire is a type of parody presented in a form of typical mainstream journalism and exhibits the elements of figurative language, especially satire [5]. News satire utilizes strongly irony and humor and mimics the characteristics of conventional news source. News satire is extremely popular on the Web.

Many research studies have been conducted in the field of natural language processing, dedicated to automatic identification of figurative languages, such as sarcasm and irony [6–8]. Yet, the automatic identification of satire is in its infancy [9]. In addition, much of the earlier studies on satire identification has been conducted for English language. In this regard, we present a machine learning based approach to satire identification in Turkish news articles. To the best of our knowledge, it is the first study in Turkish dedicated to satirical text classification based on machine learning. To examine the predictive performance of supervised learning methods on satirical text identification, we have collected 1000 satirical news articles and 1000 non-satirical news articles from two different news sources. In the presented scheme, we utilized three kinds of features to model lexical information, namely, unigrams, bigrams and trigrams. In addition, term-frequency (TF), term-presence (TP), and TF-IDF based schemes have been taken into consideration. In the classification phase, four supervised learning methods (namely, Naïve Bayes, support vector machines, logistic regression, and C4.5 algorithms) have been utilized to examine the predictive performance of different lexical representation schemes.

The rest of this paper is structured as follows: In Sect. 2, related work on the automatic identification of figurative language with emphasize on satire identification, has been presented. In Sect. 3, the methodology of the study (namely, dataset collection process, feature engineering and classification algorithms) has been introduced. In Sect. 4, experimental procedure and the empirical results of study has been presented. Finally, Sect. 5 presents the concluding remarks.

2 Related Work

This section briefly reviews the earlier work on the automatic identification of figurative languages, with emphasize on satirical text identification.

Ahmad et al. [10] presented a machine learning based approach for satire identification on news documents. In the presented scheme, binary feature weighting, TF-IDF model, bi-normal separation scheme and TF-IDF-BNS have been utilized to extract linguistic features. In the classification phase, support vector machines have been utilized as the supervised learning method. The empirical analysis on news corpus indicated that TF-IDF-BNS scheme, which integrates TF-IDF and bi-normal separation can yield higher predictive performance compared to the other conventional schemes.

In another study, Barbieri et al. [11] presented an automatic model for satire identification in Spanish advertisement news. The presented scheme utilized linguistic features, such as, frequency of a word of each tweet, ambiguity of words, part-of-speech tags, frequency of synonyms, sentiment orientations have been taken into consideration. The empirical analysis indicated that linguistic features-based representation can outperform bag-of-words based representation scheme. Similarly, Barbieri et al. [12] conducted experimental analysis on multi-lingual satirical news identification on Twitter. In this scheme, three languages (namely, English, Spanish and Italian) have been considered to empirically evaluate the predictive performance of satire identification scheme across different languages.

Moreover, Rubin et al. [13] presented a satire identification scheme based on support vector machines with enriched feature sets, containing absurdity, humor, grammar, negative affect and punctuation. The empirical analysis indicated that the ensemble of enriched feature sets can yield promising results on the identification of fake and real news documents. Another study examined the predictive performance of linguistic feature sets on automatic identification of satire for Italian political commentaries [14]. In a recent study, Perez-Rosas et al. [15] presented a machine learning based scheme for fake news identification. In the presented scheme, unigrams, bigrams, punctuation marks, psycholinguistic feature sets, readability features and syntax features have been considered. The presented scheme yields predictive performance comparable to human performance. In another study, Salas-Zarate [9] presented a machine learning based approach for satirical news identification in Spanish. In the presented scheme, psycholinguistic feature sets have been utilized to extract features from the corpus obtained from Twitter. The classification phase of the scheme has been handled by three supervised learners, namely, support vector machines, C4.5 classifier and Naïve Bayes. In another study, Ahmed et al. [16] utilized n-gram model (namely, unigram, bigram, trigram, and four-gram), term-frequency scheme and TF-IDF scheme to extract linguistic features for fake news identification.

In addition, Yang et al. [17] presented a deep learning-based approach to satirical news identification, where hierarchical neural network architecture has been utilized in conjunction with linguistic word-embeddings. More recently, Ravi and Ravi [18] utilized psycholinguistic feature sets for irony detection.

3 Methodology

This section presents the dataset collection, feature extraction process and the classification algorithms utilized in the experimental analysis.

3.1 Dataset Collection and Preprocessing

In this study, we created a dataset for the detection of the existence of satirical text in Turkish. To do so, we collected Turkish news titles from several sources where the related news is categorized manually as satirical and non-satirical. We gathered the satirical news titles from the archives of the "Zaytung" portal which is a well-known satirical newspaper in Turkey. On the other hand, we collected the non-satirical news titles from two different sources which are the Twitter account of "Milliyet" newspaper and the archives of the website www.haberler.com. Table 1 shows the amounts of document and vocabulary of the collected news based on two categories individually.

Table 1. Descriptive information regarding the dataset utilized in empirical analysis

Category	Number of documents	Total vocabulary	Total unique vocabulary
Non-satirical	500	6031	3254
Satirical	500	8564	4390
Total	1000	14595	6632

We preprocessed the collected dataset to make ready for the operations in this study. In the first step, we removed all punctuation marks, numeric characters, and extra spaces. Then, we stemmed each term by using a pure Python stemming library Snowball-stemmer which supports Turkish language. Afterwards, we removed stop words from the relevant dataset. Table 2 shows the changes in the total and the unique vocabulary amounts after preprocessing.

Table 2. Descriptive information regarding the dataset utilized in empirical analysis after preprocessing

Category	Number of documents	Total vocabulary	Total unique vocabulary
Non-satirical	500	5444	2487
Satirical	500	7955	3213
Total	1000	13399	4676

3.2 Feature Extraction Schemes

N-gram modelling is a popular feature representation scheme for language modelling and natural language processing tasks. An n-gram is a contiguous sequence of n items from a given instance of text document. In this scheme, items may be phonemes, syllables, letters, words or characters. In natural language processing tasks, word-based n-grams and character n-grams have been widely utilized. N-gram of size 1 has been referred as "unigram", N-gram of size 2 has been referred as "bigram" and N-gram of size 3 has been referred as "trigram". To model satirical news documents, we have utilized word-based n-gram models, where unigrams, bigrams and trigrams have been taken into consideration.

In the vector space model (VSM), we have considered three different schemes to represent satirical news documents, namely, term presence-based representation, term frequency-based representation and TF-IDF based representation have been considered. In term frequency-based representation, the number of occurrence of words in the documents have been counted, namely, each document has been represented by an equal length vector with the corresponding word counts. Let t denote a word, the term frequency of t in a document d is defined as TF (t, d). In term presence-based representation, presence or absence of a word in a given document has been utilized to represent text documents, such that a particular word t is represented as 1 if it is present on a particular document d and zero, otherwise.

In addition to frequency and presence-based representation schemes, term scoring schemes may be utilized to model text documents, where the importance of terms/words on a document or corpus has been represented. In this way, we have obtained nine different configurations of the dataset for the empirical analysis.

3.3 Classification Algorithms

In the classification phase, four supervised learning algorithms (namely, Naïve Bayes algorithm, support vector machines, logistic regression, and C4.5 algorithms) have been examined. The rest of this section briefly describes the supervised learning algorithms utilized in the empirical analysis.

Naïve Bayes algorithm (NB) is a probabilistic classification algorithm based on Bayes' theorem. It has a simple structure due to the assumption of conditional independence. Despite its simple structure, it can be effectively utilized in a wide range of applications, including text mining and web mining [19].

Support vector machines (SVM) are supervised learning algorithms that can be utilized to solve classification and regression problems. They can be applied effectively to classify both linear and non-linear data [20]. Support vector machines build a hyperplane in a higher dimensional space to solve classification or regression problem. The hyperplane aims to make a good separation by achieving the largest distance to the nearest training data points of classes (known as functional margin).

Logistic regression (LR) is a linear classification algorithm, which uses a linear function of a set of predictor variables to model the probability of some event's occurring [21]. Linear regression can yield good results. However, the membership values generated by linear regression cannot be always in [0–1] range, which is not an appropriate range for probabilities. In logistic regression, a linear model is constructed on the transformed target variable whilst eliminating the mentioned problems.

C4.5 is a popular decision tree algorithm [22]. To solve inherent attribute bias of ID3 algorithm, information gain ratio is used to build a decision tree for the training data set. Then, the full tree is post-pruned to solve overfitting and size problems. In the algorithm, an attribute with the highest information gain is selected. The algorithm can work properly with both continuous and discrete attributes.

4   Experimental Procedure and Results

In this section, evaluation measures, experimental procedure and experimental results have been presented.

4.1   Evaluation Measures

To evaluate the performance of classification algorithms, three different evaluation measures, namely, classification accuracy and F-measure, and area under roc curve have been considered.

Classification accuracy (ACC) is the proportion of true positives and true negatives obtained by the classification algorithm over the total number of instances as given by Eq. 1:

$$ACC = \frac{TN + TP}{TP + FP + FN + TN} \tag{1}$$

where *TN* denotes number of true negatives, *TP* denotes number of true positives, *FP* denotes number of false positives and *FN* denotes number of false negatives.

Precision (PRE) is the proportion of the true positives against the true positives and false positives as given by Eq. 2:

$$PRE = \frac{TP}{TP + FP} \tag{2}$$

Recall (REC) is the proportion of the true positives against the true positives and false negatives as given by Eq. 3:

$$REC = \frac{TP}{TP + FN} \tag{3}$$

F-measure takes values between 0 and 1. It is the harmonic mean of precision and recall as determined by Eq. 4:

$$F - measure = \frac{2 * PRE * REC}{PRE + REC} \tag{4}$$

The area under curve (AUC) is another common metric for evaluating the classifiers. It is equal to the probability that a classifier will rank a randomly chosen positive instance higher than a randomly chosen negative one. It takes on values from 0 to 1. The higher values of AUC indicate better performance of the classification algorithms.

4.2   Experimental Procedure

In the empirical analysis, 10-fold cross validation has been utilized. In this scheme, the original dataset is randomly divided into ten mutually exclusive folds. Training and testing process are repeated ten times and each part is tested and trained ten times.

The results reported in this section are the average results for 10-folds. In the empirical analysis, three different feature extraction methods (namely, term-presence based representation, term-frequency based representation and TF-IDF weighting scheme) have been taken into consideration. In addition, different sizes of n-gram (ranging from $n = 1$ to $n = 3$) have been evaluated. In this way, we obtained 9 different configurations for the dataset. In the empirical analysis, four supervised learning algorithms have been utilized to classify instances, as satirical and non-satirical.

4.3   Experimental Results

In Tables 3, 4 and 5, classification accuracy values, F-measure values and area under curve (AUC) values obtained by the compared nine representation schemes and four supervised learning algorithms have been represented, respectively. As it can be observed from the results listed in Table 3, the highest predictive performance in terms of classification accuracy (89.70%) has been achieved by unigram and term-frequency based representation and support vector machines. The second highest predictive performance has been obtained by trigram and term-frequency based representation and support vector machines, as the supervised learner. Regarding the performance of representation schemes in terms of F-measure values listed in Table 4, the highest predictive performance has been obtained with the use of unigram and term-frequency based representation and support vector machines. The second highest predictive performance has been obtained by bigram and term-frequency based representation and trigram and term-frequency based representation. For these schemes, support vector machines have been utilized as the classification algorithm. The same predictive performance patterns are also valid for the area under roc curve (AUC) values summarized in Table 4. As it can be observed from the results listed in Table 4, the highest AUC value (0.90) has been obtained by unigram and term-frequency based representation and support vector machines.

Table 3. Classification accuracy values obtained by compared representation schemes and classifiers

Representation	SVM	Logistic Regression	C4,5	Naive Bayes
Unigram, Term-frequency	**89,70**	83,95	65,53	81,47
Unigram, Term-presence	54,86	54,95	50,00	69,71
Unigram, TF-IDF	50,20	50,42	50,00	52,24
Bigram, Term-frequency	87,79	84,81	65,53	81,47
Bigram, Term-presence	55,15	58,86	50,00	69,71
Bigram, TF-IDF	50,20	52,98	50,00	52,24
Trigram, Term-frequency	*87,96*	84,58	65,53	81,47
Trigram, Term-presence	53,19	58,68	50,00	69,71
Trigram, TF-IDF	50,20	52,92	50,00	52,24

Table 4. F-measure values obtained by compared representation schemes and classifiers

Representation	SVM	Logistic Regression	C4,5	Naive Bayes
Unigram, Term-frequency	**0,89**	0,85	0,73	0,82
Unigram, Term-presence	0,69	0,34	0,67	0,73
Unigram, TF-IDF	0,67	0,03	0,67	0,68
Bigram, Term-frequency	*0,87*	0,86	0,73	0,82
Bigram, Term-presence	0,69	0,70	0,67	0,68
Bigram, TF-IDF	0,67	0,12	0,67	0,68
Trigram, Term-frequency	*0,87*	0,86	0,73	0,82
Trigram, Term-presence	0,68	0,7	0,67	0,73
Trigram, TF-IDF	0,67	0,11	0,67	0,68

Table 5. AUC values obtained by compared representation schemes and classifiers

Representation	SVM	Logistic Regression	C4,5	Naive Bayes
Unigram, Term-frequency	**0,90**	0,84	0,67	*0,88*
Unigram, Term-presence	0,55	0,55	0,50	0,76
Unigram, TF-IDF	0,50	0,50	0,50	0,51
Bigram, Term-frequency	*0,88*	0,85	0,67	*0,88*
Bigram, Term-presence	0,55	0,59	0,50	0,76
Bigram, TF-IDF	0,50	0,53	0,50	0,51
Trigram, Term-frequency	*0,88*	0,85	0,67	*0,88*
Trigram, Term-presence	0,53	0,59	0,50	0,76
Trigram, TF-IDF	0,50	0,53	0,50	0,51

To summarize the main findings of the empirical analysis, Figs. 1 and 2 present the main effect plots for accuracy values and F-measure values, respectively. Regarding the average values among the all compared schemes, bigram and term frequency-based representation, trigram and term frequency-based representation and unigram and term frequency based representation generally yield higher average classification accuracies compared to the other representation schemes.

The highest values among the nine configurations has been achieved by unigram and term-frequency based representation, with an average classification accuracy of 80.16%. Regarding the performance of classification algorithms in terms of classification accuracy values summarized in Fig. 1, Naïve Bayes yields the highest average classification accuracy (67.80%) among the compared configurations and the lowest average classification accuracy (55.17%) has been obtained by C4.5 algorithm. For the F-measure values, illustrated in Fig. 2, the same patterns that are valid for classification values in terms of different representation schemes are still valid. In terms of performance of classification algorithms, Naïve Bayes and support vector machines yield promising results, whereas logistic regression algorithm yields the lowest average F-measure values.

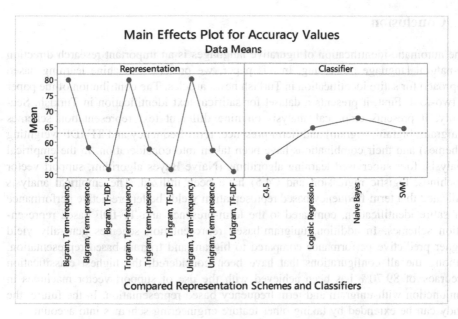

Fig. 1. The main effect plot for accuracy results

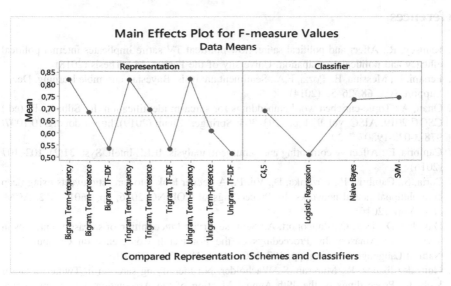

Fig. 2. The main effect plot for F-measure results

In the experimental analysis, nine different representation schemes have been examined. The experimental results indicate that term-frequency based representation yields better predictive performance for satire identification, compared to the term-presence and TF-IDF based representation schemes. In addition, unigram-based representation schemes generally yield higher predictive performance compared to bigram and trigram-based representation.

5 Conclusion

The automatic identification of figurative languages is an important research direction in natural language processing. In this paper, we present a machine learning based approach for satire identification in Turkish news articles. The contribution of the paper is two-fold. First, it presents a dataset for satirical text identification in Turkish. Secondly, it presents empirical analysis on nine different text representation schemes (unigram, bigram, trigram) and (term-presence, term-frequency and TF-IDF weighting schemes) and their combinations have been taken into consideration. In the empirical analysis, four supervised learning algorithms (Naïve Bayes algorithm, support vector machines, logistic regression and C4.5) have been utilized. The empirical analysis indicated that term-frequency based representation yields better predictive performance for satire identification, compared to the term-presence and TF-IDF based representation schemes. In addition, unigram-based representation schemes generally yield higher predictive performance compared to bigram and trigram-based representation. Among the all configurations that have been considered, the highest classification accuracy of 89.70% has been achieved with the use of support vector machines in conjunction with unigram and term-frequency based representation. In the future, the study can be extended by taking other feature engineering schemes into account.

References

1. Ramsey, R.: Affect and political satire: how political TV satire implicates internal political efficacy and political participation. University of the Pacific, MA Thesis (2018)
2. Fersini, E., Messina, E., Pozzi, F.A.: Sentiment analysis: Bayesian ensemble learning. Decis. Support Syst. **68**, 26–38 (2014)
3. Onan, A.: Topic-enriched word embeddings for sarcasm identification. In: Silhavy, R. (ed.) CSOC 2019. AISC, vol. 984, pp. 293–304. Springer, Cham (2019). https://doi.org/10.1007/978-3-030-19807-7_29
4. Cambria, E.: Affective computing and sentiment analysis. IEEE Intell. Syst. **31**(2), 102–107 (2016)
5. Poria, S., Cambria, E., Hazarika, D., Vij, P.: A deeper look into sarcastic tweets using deep convolutional neural networks. In: Proceedings of COLING 2016, pp. 1601–1612. ACM, New York (2016)
6. Davidov, D., Tsur, O., Rappoport, A.: Semi-supervised recognition of sarcastic sentences in Twitter and Amazon. In: Proceedings of the Fourteenth Conference on Computational Natural Language Learning, pp. 107–116. ACM, New York (2010)
7. Gonzalez-Ibanez, R., Muresan, S., Wacholder, N.: Identifying sarcasm in Twitter: a closer look. In: Proceedings of the 49th Annual Meeting of the Association for Computational Linguistics: Human Language Technologies, pp. 581–586. ACM, New York (2011)
8. Filatova, E.: Irony and sarcasm: corpus generation and analysis using crowdsourcing. In: Proceedings of Language Resources and Evaluation Conference, pp. 392–398. ACM, New York (2012)
9. Salas-Zarate, M., Paredes-Valverde, M.A., Rodriguez-Garcia, M.A., Valencia-Garica, R., Alor-Hernandez, G.: Automatic detection of satire in Twitter: a psycholinguistic-based approach. Knowl.-Based Syst. **128**, 20–33 (2017)

10. Ahmad, T., Akhtar, H., Chopra, A., Akhtar, M.W.: Satire detection from web documents using machine learning methods. In: Proceedings of International Conference on Soft Computing and Machine Intelligence, pp. 102–105. IEEE, New York (2014)
11. Barbieri, F., Ronzano, F., Saggion, H.: Is this tweet satirical? a computational approach for satire detection in Spanish. Procesamiento del Lenguaje Nat. **55**, 135–142 (2015)
12. Barbieri, F., Ronzano, F., Saggion, H.: Do we criticise (and laugh) in the same way? automatic detection of multi-lingual satirical news in Twitter. In: Proceedings of the Twenty-Fourth International Joint Conference on Artificial Intelligence, pp. 1215–1221. AAAI Press, New York (2015)
13. Rubin, V., Conroy, N., Chen, Y., Cornwell, S.: Fake news or truth? using satirical cues to detect potentially misleading news. In: Proceedings of the Second Workshop on Computational Approaches to Deception Detection, pp. 7–17. ACL, New York (2016)
14. Delmonte, R., Stingo, M.: Detecting satire in italian political commentaries. In: Nguyen, N.-T., Manolopoulos, Y., Iliadis, L., Trawiński, B. (eds.) ICCCI 2016. LNCS (LNAI), vol. 9876, pp. 68–77. Springer, Cham (2016). https://doi.org/10.1007/978-3-319-45246-3_7
15. Perez-Rosas, V., Kleinberg, B., Lefevre, A., Mihalcea, R.: Automatic detection of fake news. arXiv preprint arXiv:1708.07104 (2017)
16. Ahmed, H., Traore, I., Saad, S.: Detection of online fake news using n-gram analysis and machine learning techniques. In: Traore, I., Woungang, I., Awad, A. (eds.) ISDDC 2017. LNCS, vol. 10618, pp. 127–138. Springer, Cham (2017). https://doi.org/10.1007/978-3-319-69155-8_9
17. Yang, F., Mukherjee, A., Dragut, E.: Satirical news detection and analysis using attention mechanism and linguistic features. arXiv preprint arXiv:1709.01189 (2017)
18. Ravi, K., Ravi, V.: Irony detection using neural network language model, psycholinguistic features and text mining. In: Proceedings of IEEE 17th International Conference on Cognitive Informatics and Cognitive Computing, pp. 254–260. IEEE, New York (2018)
19. Onan, A.: Classifier and feature set ensembles for web page classification. J. Inf. Sci. **42**(2), 150–165 (2016)
20. Cortes, C., Vapnik, V.: Support-vector networks. Mach. Learn. **20**(3), 273–297 (1995)
21. Kantardzic, M.: Data Mining: Concepts, Models, Methods and Algorithms. Wiley, Hoboken (2011)
22. Gehrke, J.: The Handbook of Data Mining. Lawrence Erlbaum Associates, Chicago (2003)

Big Data Innovation and Applications

Committee of the SGTM Neural-Like Structures with Extended Inputs for Predictive Analytics in Insurance

Roman Tkachenko[1] , Ivan Izonin[1]([⊠]) , Michal Greguš ml.[2] ,
Pavlo Tkachenko[3], and Ivanna Dronyuk[1]

[1] Lviv Polytechnic National University, Lviv, Ukraine
roman.tkachenko@gmail.com, ivanizonin@gmail.com,
ivanna.droniuk@gmail.com
[2] Comenius University in Bratislava, Bratislava, Slovakia
Michal.Gregusml@fm.uniba.sk
[3] IT STEP University, Lviv, Ukraine
pavlo.tkachenko@gmail.com

Abstract. In this paper, we propose a new committee-based method for insurance data analytics problem. The main idea of the method is to increase the prediction task's accuracy. The division of the dataset into segments using dichotomy approach are performed in accordance with the developed algorithm. Both the Kolmogorov-Gabor polynomial and the neural-like structures of the Successive Geometric Transformation Model are proposed for the division and for the prediction procedures. Such a combination provides high accuracy with satisfactory time characteristics of the training procedure. A number of experiments are carried out on the accuracy of the method and the speed of the training procedure. The highest accuracy of the developed method in comparison with the existing ones is established. The main advantages and disadvantages of the proposed method are outlined. The proposed approach can be used to efficiently solve regression and classification tasks for insurance.

Keywords: Prediction task · Committee · Segmented regression ·
Kolmogorov-Gabor polynomial · Insurance costs · SGTM neural-like structure

1 Introduction

The insurance data analytics is an important task today [1]. In the era of Big Data, Predictive Analytics in insurance is a complex task due to the number of issues:

- the need to process huge data sets [2];
- the complex, parametric relationships between the large number of independent variables of each vector from a set [3];
- the need to find a compromise between the accuracy and speed of the constructed model for solving one or another applied task [4].

Among a large number of existing methods developed to solve this task, the linear or logistic regression techniques in most cases remain the appropriate tools for

© Springer Nature Switzerland AG 2019
M. Younas et al. (Eds.): Innovate-Data 2019, CCIS 1054, pp. 121–132, 2019.
https://doi.org/10.1007/978-3-030-27355-2_9

constructing an accurately predicted model. They are especially effective when using a segment-based approach. Building a separate model for each individual segment from a single dataset should increase the accuracy of prediction. However, this approach increases the time spent on processing each individual data segment, which is a significant drawback in modern online information processing systems [5].

Nevertheless, the ability to use recent research in the field of computational intelligence allows using new, more efficient tools for processing each individual segment from the dataset that will improve the accuracy of the entire model. And the use of parallel computing [6] for the processing of all segments will reduce the time of processing from the large dataset.

2 Review and Analysis of Existing Methods

Modern Predictive Analytics technics that are based on Data mining [7] and Data-stream mining methods [8] allows seeing hidden patterns in datasets that people are not able to see. However, there are a lot of problems in Big Data intelligent analysis tasks, such as enormous data sets, correct selection of the artificial intelligence method, and its configuration for the data processing, the risks of its overfitting or underfitting, low data processing speed, etc. This leads to the search for new ones, simpler and more efficient solutions. One of these may be the Segment-based approach using artificial intelligence tools.

In the [9] it is investigating the approximation properties of the segmented multiple regression method. The estimation of regression parameters according to the method takes place using the least squares. In the [10] it is considered the possibility of using segmented polynomial regression techniques for application in neurobiology. The search for the coefficients for the piecewise polynomials is also based on the use of the least squares. In addition, authors of this paper attempts are made to apply the weighted least squares method to find the coefficients of the piecewise polynomials, that used by each individual segment. Both methods for searching of the polynomial coefficients are rather time-consuming, and considering both: the need to process large volumes of data and the possibility of usage high degree polynomials. The methods described above impose a number of limitations on their practical application.

The paper [11] analyzes the asymptotic properties of segmented regression models, where the segments are forming based on clustering. The evaluation of the model parameters is based on the method of maximum likelihood and its hybrids. In [4] authors are describing the simplified Cluster-Regression Approximation model, which shows satisfactory results according to the criteria, proposed by the authors. However, the main disadvantage of both methods is the need for a clustering procedure. First of all, this requires proper selection of both: the clustering method and the number of required clusters, which is a completely nontrivial task [5]. In addition, these procedures require a lot of time resources [12] and therefore such methods cannot always be used for the Big Data processing tasks.

The author of [13] suggests using a piecewise step-by-step approach based on neural-like structures of the Successive Geometric Transformations Model (SGTM) for solving the classification task. The procedures for separating the unbalanced data sample according to the method [13] as in the previous group of methods require the application of a certain clustering algorithm (k-means). This fact imposes a number of limitations on the practical application of the method in the applied online processing systems.

However, the use of high-speed SGTM neural-like structure, which is used a greedy-based non-iterative training procedure, can significantly reduce the processing time. Mathematical apparatus and detailed description of the training and application procedures, as well as formation of the activation function as some approximation dependence of this computational intelligence tools, are given in [14].

A series of experimental investigations for applying the Kolmogorov-Gabor Polynomial and SGTM neural-like structure for solving the regression task during the processing of the entire data sample is given in [15]. In these works, the polynomial is used as an effective tool for constructing a nonlinear model, and the SGTM neural-like structure is used for the fast and accurate procedure of searching the coefficients for this polynomial. Based on the series of investigations, in [16] it proved the effectiveness of the combined use of both these tools for solving the prediction of medical insurance costs.

The paper [17] describes a Segmented-based Multiple Regression Method constructed using only the SGTM neural-like structure. The method shows the highest results regarding the speed of the training algorithm and good results in prediction accuracy.

However, as discussed in [18] the additional application of Kolmogorov-Gabor polynomial can improve the accuracy of the constructed model.

Based on above analysis, in this paper authors propose a new Committee based on consistent use of the Kolmogorov-Gabor polynomial and SGTM neural-like structures. Given the high accuracy of the polynomial approximation, as well as the high speed of the neural-like structure, the proposed approach must provide good results for solving the stated task.

3 Proposed Committee of the SGTM Neural-Like Structures with Extended Inputs

We have considered the multiple regression task for the case of a large volume of data processing. The task is to predict dependent variable y based on the set of independent ones $x_1, \ldots, x_n$ by dividing the sample of large-scale observation into parts (segments). We propose a new, segmented regression method for solving this task based on the combining usage of the Kolmogorov-Gabor Polynomial and non-iterative SGTM neural-like structure of the linear type. The combination of both of these tools provides effective results in terms of the method accuracy [18] and the speed [14] of the training algorithm.

The main objective of the proposed approach is to improve prediction accuracy. For this purpose, it uses the a divide strategy. We divide the dataset into parts by comparing the predicted values of the dependent variable y_i^{pred} to its mean value $y_i^{average}$ in the sample of the current step i (dichotomy approach). As the steps of the division increase, the number of subsets doubles. Each of them uses its SGTM neural-like structure with previous processing of input data based on Kolmogorov-Gabor Polynomial.

A block diagram of the sample division for the formation of one step of the division is demonstrated on Fig. 1. Each sorting blocks are work as Sorting Block 1. It means that each block is divided dataset for two part according to it average value $y_i^{average}$. However, in the case of processing larger volumes of data, such a division can be continued. It should be noted that such an approach allows using of parallel-distributed computing since data samples do not overlap. This will accelerate training procedures for solving the different task [3, 19–21], especially for large volumes of data processing [2].

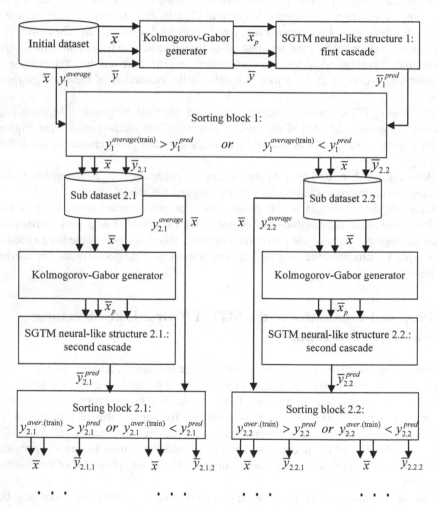

Fig. 1. Block diagram of the proposed segmented regression method

The main difference of the developed segmented regression method from the method [17] is that Kolmogorov-Gabor Polynomial of the second degree [15] is additionally used for the division and prediction procedures at each step of the method:

$$Y(x_1, \ldots, x_n) = \theta_0 + \sum_{i=1}^{n} \theta_i x_i + \sum_{i=1}^{n} \sum_{j=i}^{n} \theta_{i,j} x_i x_j. \tag{1}$$

It increases approximation properties of the method. Moreover, the use of a second-order polynomial provides satisfactory time delays for the search of its coefficients θ_0, θ_i, $\theta_{i,j}$ using the fast SGTM neural-like structure [16].

4 Modeling and Results

The modelling of the proposed method was performed on a data sample from [22]. This is the task of individual insurance cost prediction in the USA. The matrix of training data contained 1070 vectors, and the matrix of the dimensional of the test data matrix was 268 vectors. The detailed characteristics of the initial dataset are given in Table 1.

Table 1. Initial dataset.

Data vector	Characteristics
Contractor's age	Mean: 39.2; min: 18 years; max: 64 years
Sex	Male: 676; female: 662
Body mass index, kg/m^2	Min: 15.96; max: 53.13; mean: 30.66
Number of children	Min: 0; max: 5; mean: 1.095
Smoking	Smokers: 1064; no-smokers: 274
Residential area	Northwest: 325; southeast: 364;
	Northeast: 324; southwest: 325

In order to use the computational intelligence methods, the selected data sample has been modified (using binary system) and its general view is given in Table 2 [16].

As a result of the modification, instead of the 6, 11 independent variables participated in the simulation process for each input data vector (columns *Sex, Smoking and Residential* area were divided to 2, 2 and 4 columns accordingly). Figure 2 presents the Dataset visualization.

Modelling results were evaluated using 3 indicators:

- Mean absolute percentage error (MAPE):

$$MAPE = \frac{100}{n} \sum_{i=1}^{n} \left| \frac{y_i^{true} - y_i^{pred}}{y_i^{true}} \right|, \tag{2}$$

Table 2. Prepared dataset.

Data vector	Characteristics
Contractor's age	Mean: 39.2; min: 18 years; max: 64 years
Male	Male: 676
Female	Female: 662
Body mass index, kg/m^2	Min: 15.96; max: 53.13; mean: 30.66
Number of children	Min: 0; max: 5; mean: 1.095
Smoking: male	Smokers: 1064; no-smokers: 274
Smoking: female	Smokers: 274; no-smokers: 1064
Residential area A	Northwest: 325
Residential area B	Southeast: 364
Residential area C	Northeast: 324
Residential area D	Southwest: 325

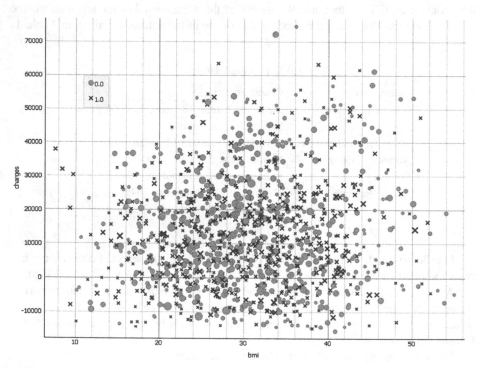

Fig. 2. Visualization of the data sample: the x-axis represents the body mass index; the y-axis represents the level of insurance medical costs. The blue circles mean the women and the red cross symbolizes men. The figure's size reflects the number of children: the larger size the larger the number of children in the family. (Color figure online)

– Root-mean-square error (RMSE):

$$RMSE = \sqrt{\frac{\sum_{i=1}^{n} \left(y_i^{pred} - y_i^{true}\right)^2}{n}}, \tag{3}$$

– Training time, in seconds.

where:

y_i^{true} is the real value;

y_i^{pred} is the predicted value;

n is the dimension of the dataset (training or test);

Since it is about estimating an error on several samples, we used their weighed value.

As indicated in [8, 9] the application of the second-degree Kolmogorov-Gabor polynomial increases the number of inputs, and in our case, instead of 11 input attributes (Table 2), we process 77 for each input vector. That is why the topology of SGTM neural-like structures at each division step will contain 77 neurons in the input and hidden layers and 1 neuron in the output layer.

The results of the proposed method are presented in Table 3. It contains the results of conducted experiments for one step of division in both the training and the test mode. The results of the method, according to MAPE show very close values in both modes.

Table 3. Proposed method's results.

Sample	Parameters		
	dim (training sample), vectors	dim (test sample), vectors	MAPE, %
Training mode			
First subsample	731	186	28.692
Second subsample	339	82	31.367
Weighted value	1070	268	29.433
Test mode			
First subsample	731	186	33.749
Second subsample	339	82	18.957
Weighted value	1070	268	29.223

Figure 3 shows the scatter plots (using Orange software [23]) of the real data to the predicted values for both obtained sub-samples.

As can be seen in Fig. 3, the performance of our method shows the best indicators in the first sub-sample than in the second. However, a weighted value, based on both sub-samples, provides satisfactory results.

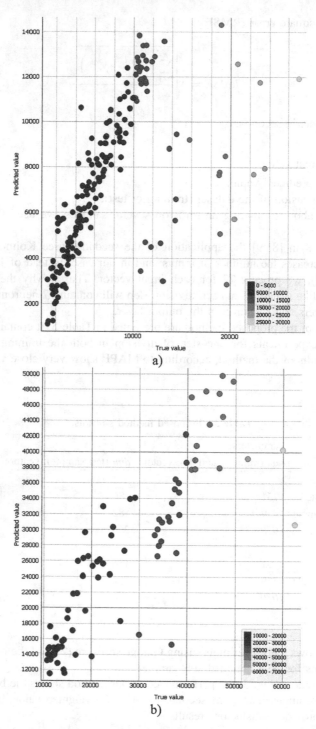

Fig. 3. Obtained results based on: (a) first subsample; (b) second subsample

5 Comparison and Discussion

In order to compare the results of our method, existing regression methods were used, in particular:

- Stochastic Gradient Descent regressor;
- Multi-layer Perceptron;
- General Regression Neural Network;
- SGTM neural-like structure;
- The piecewise linear approach using SGTM neural-like structure (Method from [17]);
- The predictor based on Kolmogorov-Gabor polynomial and SGTM neural-like structure [16].

It should be noted that all known methods worked overall sample, where the ratio of training and test parts was 80 and 20% respectively. The results of numerical simulation of all investigated methods are shown in Table 4. As can be seen from Table 4, the best results for all indicators are obtained using the proposed model. A similar approach, however, without the use of inputs polynomial extension [17], provides satisfactory results for the accuracy of work (inferior to the proposed one less than 2% (MAPE)). However, the calculated RMSE values show a rather large difference, and given that it is a question of money, this is a significant loss of funds [24, 25].

If talking about training time, in most cases, it plays an important role for design real systems based on computational intelligence tools.

This task is even more complicated in the case of the large volumes of data processing, where time delays will increase significantly. That is why in this paper experimentally determined the duration of training procedures of all investigated methods compare to the proposed one. It should be noted that the duration of the training procedures of the developed method was calculated as the sum of the length of training procedures at each step of the division. Here, the time of the training procedure at each separate step of the division was calculated as the maximum value among the two possible in this step. This approach is due to the possibilities of parallel distributed computing because sub-samples of each individual step do not overlap. The results of this study are shown in Fig. 4.

As can be seen in Fig. 4, the method based on the Stochastic Gradient Descent shows the best results. However, according to Table 4 the accuracy of its work is the worst. A Multilayer Perceptron with unsatisfactory accuracy of work in 52% also shows the worst result due to the working time of the training process.

Considering this, these methods are not recommended for use when designing smart health insurance management systems. Given the significant increase in the size of the input data of the developed model (from 11 to 77 inputs), it shows satisfactory results regarding the duration of the training procedure.

Table 4. The comparison of the all method's predicted results (test mode).

#	Method	Parameters	Accuracy indicators	
			MAPE, %	RMSE
1	Stochastic Gradient Descent (SGD regressor)	Loss = 'squared loss', penalty = 'l2', alpha = 0.0001	63.458	6586.878
2	Multi-layer Perceptron (MLP)	11 inputs, 11 neurons in the hidden layer, one output, 200 epochs	52.929	6393.225
3	Common SGTM neural-like structure (SGTM NLS)	11 inputs, 11 neurons in the hidden layer, one output	41.919	6095.301
4	General Regression Neural Network (GRNN)	s = 0.1 (sigma = s, (s ∈ [0.1, 1.5], Δ = 0.1),	31.913	5905.643
5	Piecewise Linear Approach using SGTM neural-like structure [17]	One division step, 11 inputs, 11 neurons in the hidden layer, one output	30.604	5107.095
6	Predictor based on the Kolmogorov-Gabor polynomial & SGTM neural-like structure [16]	77 inputs, 77 neurons in the hidden layer, one output	30.101	5030.181
7	**Committee of the SGTM Neural-like Structures with Extended Inputs (using Kolmogorov-Gabor Polynomial)**	**One division step, 77 inputs, 77 neurons in the hidden layer, one output.**	**29.223**	**4873.173**

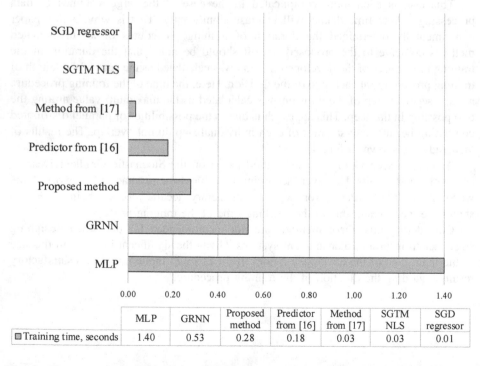

	MLP	GRNN	Proposed method	Predictor from [16]	Method from [17]	SGTM NLS	SGD regressor
▢ Training time, seconds	1.40	0.53	0.28	0.18	0.03	0.03	0.01

Fig. 4. The comparison of the training time for all methods. The x-axis shows the training time, seconds

6 Conclusion

We present the new non-iterative approach for insurance costs prediction task. It is built on committee of the SGTM neural-like structure with extended inputs. The algorithmic realization of the method is described, its structural scheme is presented. We solved the medical insurance costs prediction task. The main purpose of the method was to improve the prediction results. The developed method using only one-step of division (only two data clusters) shows high accuracy.

The effectiveness of the proposed approach is confirmed by comparison with known methods. The proposed approach provides the highest accuracy of work with satisfactory characteristics of the duration of the training procedure.

The proposed approach, based on the non-iterative machine learning algorithm, can be used to solve the classification and regression tasks in the conditions of a large volume of data processing for insurance data analytics.

References

1. Wu, D.: A big data analytics framework for forecasting rare customer complaints: a use case of predicting MA members' complaints to CMS. In: 2017 IEEE International Conference on Big Data (Big Data), pp. 3965–3967 (2017)
2. Shakhovska, N.B., Bolubash, Y.J., Veres, O.M.: Big data federated repository model. In: The Experience of Designing and Application of CAD Systems in Microelectronics, pp. 382–384 (2015)
3. Babichev, S., Lytvynenko, V., Škvor, J., Korobchynskyi, M., Voronenko, M.: Information technology of gene expression profiles processing for purpose of gene regulatory networks reconstruction. In: 2018 IEEE Second International Conference on Data Stream Mining Processing (DSMP), pp. 336–341 (2018)
4. Molnár, E., Molnár, R., Kryvinska, N., Greguš, M.: Web intelligence in practice. J. Serv. Sci. Res. 6(1), 149–172 (2014)
5. Kaczor, S., Kryvinska, N.: "It is all about Services - Fundamentals, Drivers, and Business Models", the society of service science. J. Serv. Sci. Res. 5(2), 125–154 (2013)
6. Tsmots, I., Skorokhoda, O., Rabyk, V.: Structure and software model of a parallel-vertical multi-input adder for FPGA implementation. In: Proceedings of the 11th International Scientific and Technical Conference Computer Sciences and Information Technologies - CSIT 2016, pp. 158–160 (2016)
7. Setlak, G., Bodyanskiy, Y., Vynokurova, O., Pliss, I.: Deep evolving GMDH-SVM-neural network and its learning for data mining tasks. In: Proceedings of the 2016 Federated Conference on Computer Science and Information Systems, FedCSIS 2016, pp. 141–145 (2016)
8. Bodyanskiy, Y., Vynokurova, O., Szymanski, Z., Kobylin, I., Kobylin, O.: Adaptive robust models for identification of nonstationary systems in data stream mining tasks. In: Proceedings of the 2016 IEEE 1st International Conference on Data Stream Mining and Processing, DSMP 2016, pp. 263–268 (2016)
9. Kim, J., Kim, H.-J.: Asymptotic results in segmented multiple regression. J. Multivar. Anal. 99(9), 2016–2038 (2008)
10. Weiershäuser, A.: Piecewise polynomial regression with fractional residuals for the analysis of calcium imaging data. Ph.D. Thesis, University of Konstanz (2012)

11. Robinson, R.A.: Asymptotics and confidence estimation in segmented regression models. Ph.D. Thesis, University of Louisville (2012)
12. Hu, Z., Bodyanskiy, Y., Tyshchenko, O.K.: Self-learning procedures for a kernel fuzzy clustering system. Adv. Intell. Syst. Comput. **754**, 487–497 (2019)
13. Doroshenko, A.: Piecewise-linear approach to classification based on geometrical transformation model for imbalanced dataset. In: 2018 IEEE Second International Conference on Data Stream Mining Processing (DSMP), pp. 231–235 (2018)
14. Tkachenko, R., Izonin, I.: Model and principles for the implementation of neural-like structures based on geometric data transformations. In: Hu, Z., Petoukhov, S., Dychka, I., He, M. (eds.) ICCSEEA 2018. AISC, vol. 754, pp. 578–587. Springer, Cham (2019). https://doi.org/10.1007/978-3-319-91008-6_58
15. Vitynskyi, P., et al.: Hybridization of the SGTM neural-like structure through inputs polynomial extension. In: 2018 IEEE Second International Conference on Data Stream Mining & Processing (DSMP), Lviv, Ukraine, pp. 386–391 (2018)
16. Tkachenko, R., et al.: Development of the non-iterative supervised learning predictor based on the ito decomposition and SGTM neural-like structure for managing medical insurance costs. Data **3**(4), 46 (2018)
17. Tkachenko, R., et al: Piecewise-linear approach for medical insurance costs prediction using SGTM neural-like structure. In: Proceedings of the 1st International Workshop on Informatics & Data-Driven Medicine (IDDM 2018), Lviv, Ukraine, vol. 2255, pp. 170–179 (2018)
18. Gregus, M., Kryvinska, N.: Service Orientation of Enterprises - Aspects, Dimensions, Technologies. Comenius University in Bratislava (2015). ISBN: 9788022339780
19. Kazarian, A., Teslyuk, V., Tsmots, I., Mashevska, M.: Units and structure of automated 'smart' house control system using machine learning algorithms. In: 2017 14th International Conference the Experience of Designing and Application of CAD Systems in Microelectronics (CADSM), pp. 364–366 (2017)
20. Lytvyn, V., Vysotska, V., Veres, O., Rishnyak, I., Rishnyak, H.: The risk management modelling in multi project environment. In: 2017 12th International Scientific and Technical Conference on Computer Sciences and Information Technologies, vol. 1, pp. 32–35 (2017)
21. Korjagin, S., Zarov, V., Klachek, P., Polupan, K.: Creation of intellectual decision-making systems in industry based on hybrid computing methods. In: IOP Conference Series: Materials Science and Engineering, Conference 1, vol. 497, pp. 1–8 (2019)
22. Medical Cost Personal Datasets. https://www.kaggle.com/mirichoi0218/insurance. Accessed 08 Dec 2018
23. Demšar, J., et al.: Orange: data mining toolbox in Python. J. Mach. Learn. Res. **14**, 2349–2353 (2013)
24. Kryvinska, N., Gregus, N.: SOA and its Business Value in Requirements, Features, Practices and Methodologies. Comenius University in Bratislava (2004). ISBN: 9788022337649
25. Kryvinska, N.: "Building consistent formal specification for the service enterprise agility foundation", the society of service science. J. Serv. Sci. Res. **4**(2), 235–269 (2012)

Game Analytics on Free to Play

Robert Flunger, Andreas Mladenow$^{(\boxtimes)}$, and Christine Strauss

University of Vienna, Oskar-Morgenstern-Platz 1, 1090 Vienna, Austria
{flungerr58, andreas.mladenow,
christine.strauss}@univie.ac.at

Abstract. This contribution refers to a prosperous digital business segment, i.e. gaming business. We identify game analytics aspects with a focus on analytical and predictive models for free to play (F2P) business models. The main areas of interest in this category are player churn prediction and customer lifetime value (CLV) prediction. Additionally, our study discusses why small and medium-sized developers should use game analytic tools. Based on a literature review we analyse three categories regarding game analytics: the first one describes motivations for small and medium sized game developers to use game analytic tools; the second one contains six studies discussing churn prediction models in F2P games; the third one contains four studies on prediction of customers' lifetime value.

Keywords: Free to play · Online gaming industry · Game analytics · Big data analytics · Business model · Business intelligence · Decision making · Customer Lifetime Value

1 Introduction

Over the last few years, markets enabled through information and communication technologies (ICT) have grown at a tremendous pace. With this growth many novel business models and opportunities arose, enabled through innovations such as high-speed internet and high quality mobile devices [2, 13, 21, 23, 31]. As one of the fastest growing industries in the world, the online gaming industry probably stands out the most [1, 28]. For researchers, this market is of particular interest, as it is the first truly digital native industry with the inherent potential to disrupt traditional business concepts [14]. The industry is characterized by a high degree of innovation, such as digital distribution, downloadable content, independent game development, early access titles and the free-to-play (F2P) business model [18, 27]. Especially the latter is relevant, as it has become the most successful monetization model in games [17].

The emergence of F2P approaches led to the erosion of traditional pay-to-play (P2P) business models. For example, the successful online role-playing game World of Warcraft which requires a monthly subscription to play, is reported to have dropped about one third of their subscribers from 2010 to 2013, as gamers trended towards free alternatives [6]. Another example is Valve's game Team Fortress 2, which first launched as retail game in 2007. In 2012 the game was made completely free-to-play, with an in-game shop selling virtual goods. Doing so increased the revenue of the game by a factor of twelve [19].

© Springer Nature Switzerland AG 2019
M. Younas et al. (Eds.): Innovate-Data 2019, CCIS 1054, pp. 133–141, 2019.
https://doi.org/10.1007/978-3-030-27355-2_10

Overall, the global video game market was estimated to have a revenue of $78.6 billion in 2017 and is expected to grow to $90.1 billion in 2020. Revenue of F2P games accounted for 17.1$ billion in June 2016, compared to 2.8$ of P2P revenue. Thus, comparing F2P to pay-to-play (P2P), F2P produces about 85% of total revenue. Most of this revenue is produced in Asia, especially China, Japan and South Korea [3]. Monetization is not upfront anymore but happens continuously through selling virtual goods that enhance the gaming experience. However, the conversion rate from non-paying to paying players is reported to be extremely low, rarely over 5% [29].

In the online gaming industry, F2P games massively increased their market share in a short amount of time, causing more and more businesses to rethink their (subscription-based) models [22]. It is noteworthy though that such a switch was rarely successful, indicating that developers do not understand the intricacies of the new business model to its full extent [5].

Accordingly it seems promising for research and practice to analyze all form of data that pertains to a F2P business model. In this regard, game analytics has recently emerged as a new field for data mining in the online gaming industry [8–10]. Against this background, this paper investigates game analytics aspects of the F2P business model based on a literature analysis. Hence, Sect. 2 pinpoints game analytics using F2P and Sect. 3 finishes with a short conclusion.

2 Aspects of Game Analytics

F2P games are available to play free of charge in their basic form. However, the gameplay is restricted in some ways. For example, by time constraints or unavailability of certain areas or actions in the game. These additional parts can be unlocked by paying a fee. Some games also include advertising or offer optional premium subscriptions. The most common model nowadays is in-game purchases in the form of a broad variety of virtual items that enhance the gaming experience [5, 17, 18, 25]. Due to its characteristics, F2P can thus be regarded as variation of the freemium business model [18]. Because of its nature of selling virtual items for low amounts of money, the F2P model is also referred to as "microtransactions revenue model" [4, 11].

For the methodological part of this paper a comprehensive literature review has been conducted. For the review five databases were inquired to find publications on the topic of the free-to- play model in a business context. These are (1) ACM Digital Library, (2) EBSCOhost, (3) IEEE Explore, (4) SpringerLink and (5) Wiley Online Library. This paper investigates the key results of eleven scientific papers based on the conducted literature review. Covering aspects of game analytics for the F2P business model the papers were categorized into the following three categories: Motivation to use Game Analytics, Customer Lifetime Value Prediction, and Churn Prediction and are depicted in Table 1 in chronological order.

Furthermore, Table 2 displays the frequency distribution of game environments for game analytics in the literature displayed in Table 1. In the following subsections the three categories will be discussed in detail.

Table 1. Aspects of game analytics.

Author(s)	Method		Category		
	Quantitative	Qualitative	Motivation to use Game Analytics	CLV prediction	Churn prediction
Hadiji et al. (2014) [10]	x				x
Runge et al. (2014) [26]	x				x
Hanner and Zarnekow (2015) [12]	x			x	
Koskenvoima/Mäntimäki (2015) [15]		x	x		
Sifa et al. (2015) [30]	x			x	
Lee et al. (2016) [16]	x				x
Perianez et al. (2016) [24]	x				x
Voigt and Hinz (2016) [32]	x			x	
Milosevic et al. (2017) [20]	x				x
Demediuk et al. (2018) [7]	x				x
Drachen et al. (2018) [8]	x			x	x

Table 2. Distribution of game environments for game analytics

Category	PC	Social network game	Mobile
Motivation to use Game Analytics	1	0	1
Customer Lifetime Value Prediction	1	0	4
Churn prediction	1	4	5

2.1 Motivation to Use Game Analytic Tools for SME

A qualitative study based on interviews with game developers revealed that small and medium-sized enterprises (SME) apply game analytics tools for two purposes:

(i) analytics as a communication tool and
(ii) analytics as a decision support tool.

The first one is about the necessity of reporting to investors and publishers in a comprehensive way by using for example key performance indicators. Hereby, retention rate was considered the most important metric. Furthermore, measuring CLV was deemed important as it helps to understand the extent to which the costs related to customers' acquisition are covered [15].

2.2 Churn Prediction

Churn prediction in a gaming context denotes the process of detecting and defining players of a game, who will leave the game for good at a certain future point of time. Those players, who leave the service are called "churners", and the ratio of these over non-churning players represents the so-called "churn rate". Particularly in an F2P environment the prediction of churners is an important task, as retaining players is generally considered less expensive than recruiting new ones [10]. By being able to predict when a player is about to leave, developers can adjust the gameplay experience on a more individual level and thus prolong the user's lifetime [26]. It is to mention, that all studies in this section (except one) refer to casual mobile games.

For churn prediction, a system needs to be implemented that is able to differentiate between players in a reliable way. The most common approach is a simple binary classification of players, i.e. churners and returning players [10, 16, 20, 26]. Advanced approaches are based on machine learning algorithms, which can be trained by using game datasets. Popular algorithms are neural networks, logistic regression, decision trees, support vector machines, and Markov models. However, some approaches have also been criticized. While a binary classification is intuitive and relatively easy to implement, the results are rather limited. It cannot properly process temporal information and is inflexible in predicting exact churn times and probabilities. Thus, a model based on survival analysis was suggested as this allows to produce utility functions with clear probabilities of player churn at any given point in time [24].

Hadiji et al. (2014) used binary classification, introducing a 7-day window after which a player was classified as churned [10]. They developed a prediction model and used it on an experimental dataset of four games. Hereby, results showed that the most important indicators to predict player churn were number of play sessions, number of days since sign-up, average time between play sessions and current absence time. Additionally, Runge et al. (2014) analysed in detail so-called high-value players in two games [26]. In their study high-value players are the top 10% of paying players over the last 90 days before the study took place. Features important to churn prediction were not discussed in this paper. However, the authors tested how churning players may be effectively manipulated. Thus, they sent substantial amounts of in-game currency to players who were predicted to churn or have recently churned. It turned out that this action had no significant impact on the overall churn rate. For this reason, they recommended to cross-link churning players to other games in the developer's portfolio, rather than trying to retain them in the game they are about to leave.

Furthermore, Perianez et al. (2016) found that high-value players can generally be regarded as churned when they did not play for more than 10 consecutive days [24]. As most important variables to predict churn they listed the amount of last purchase, days since last purchase and the user's level in the game. In addition, Lee et al. (2016) found that the number of purchases and the number of times attending the player's guild were the most important predictors for churning [16]. Also, the amount of virtual currency left after the last logout showed to be a relevant indicator. Especially attending the guild is an interesting element, as it implies the relevance of social factors in the player's decision to stay with a game.

Another study discussed early churn and proposed a personalized targeting strategy to retain players. Early churn denotes the first day a user starts playing the game, as it is pointed out that this period has the highest churn rate overall. After developing the prediction model, they sent push notifications to players that were likely to churn soon. The notifications had two different occurrences: they either explained game features the player used a lot in more detail, or they presented rather unexplored core features.

The goal of both activities was to re-attract and boost the player's interest in the game and motivate them to continue to play. It turned out that the first approach would not lead to a higher return rate compared to sending no notifications. Furthermore, the second approach of presenting unexplored features would significantly improve retention. This is believed to be the case as players are invited to explore something new to them, while in the first case they already may have decided that they do not like the game enough to play it [20].

Concerning the exact definition of player churn, some variants can be observed. Especially in the case of binary modeling it is important to set a certain cut-off point at which a player is classified as churned. While Lee et al. (2016), Runge et al. (2014), and Milošević et al. (2017) classify players as "churned" when they did not play for 14 days, Hadiji et al. (2014) use a 7-day-window, while Perianez et al. (2016) conclude that 10 days is a viable point of classification [11, 15, 20, 24, 26].

Finally, one study in the sample discussed a different environment, namely the popular desktop F2P game "League of Legends". In this study, survival analysis was employed to predict when players churn from the game. It is to note that "League of Legends" does not take place in a persistent game world. Instead it is a competitive player versus player (PvP) game in separate matches. For instance, the analysis showed that the time span between the matches is the strongest predictor for player churn. It indicates, the longer the time between matches, the lower the probability for players to return. In addition, it was found that a longer duration of a single match would lead to a longer time between subsequent matches. This is in line with the findings of the aforementioned studies on casual games, where the time between play sessions was a significant predictor for player churn [7].

2.3 Customer Lifetime Value Prediction

Another motive in terms of game analytics was the use of CLV theory to predict and discuss purchase behavior. According to CLV theory, purchase behavior consists of three steps, which are *(i)* customer acquisition, *(ii)* retention, and *(iii)* expansion. Acquisition is the period from installing a game until the occurrence of the first purchase. With the conversion to paying customers the retention phase starts, which ultimately leads to an expansion in the form of efficient monetization and thus a high customer value [12].

It was shown that users who do not start playing a game on the first day of installing it have a very low probability of converting to paying users. However, it turned out that retention rate increases with repeated purchases. Especially after the third repeated purchase within the specified time period the probability of purchasing again was very high. Also, the average amount spent per purchase would generally increase with following purchases. While on the first purchase, the mean volume was rather low it rose significantly from the second purchase onwards. This was explained by users purchasing small packages first to try them out, and when gaining trust in the product they would continue to choose more expensive options. Basically, this means that users who are not attracted by the game at first are likely to never change their perception. Thus, it is important to convince them and make a good impression right from the start. As for retaining users it was suggested to keep enjoyment levels relatively high at the beginning of their lifetime, to motivate them to repeated purchasing. Also, in terms of marketing it was implied that developers have to be careful about which items are advertised to whom, as advertising expensive packages to new users may turn them off, as they would perceive the purchase as too much of a risk. Thus, a dynamic presentation of potentially appropriate items per stage in the customer's lifetime cycle is suggested [12].

Voigt and Hinz (2016) investigated how the time until a user makes her/his initial purchase influences the CLV [32]. It turned out that the correlation is negative, meaning that the longer it takes until the user performs the first purchase, the lower the expected CLV. Additionally, results show that the amount of money spent on the first purchase has a positive correlation with future CLV. This implies that the respective customers are likely to spend a high amount of money on future purchases as well.

Interestingly, it was also found that payment methods hold a significant value in predicting future CLV. Customers paying with credit card showed a much higher remaining CLV than those using other methods. The authors noted that up to the date of their study, literature on effect of payment models in digital businesses has been very scarce [2] Thus, this might be a research avenue worth exploring in the future.

Furthermore, the findings of Hanner and Zarnekow (2015) that the average purchase amount increases with subsequent purchases, was also confirmed by Voigt and Hinz [12, 32]. The latter showed in their prediction model that with subsequent purchases the average amount of money spent per purchase rises. However, the higher this amount got, the less likely it was to change any further. This is an important finding as identifying heavy spenders as early as possible allows to target them efficiently with marketing incentives. For example, high-value customers could be served with benefits such as exclusive content and faster response time to service inquiries [32]. Another CLV prediction model was established by Drachen et al. (2018) [8]. Based on a casual mobile game they classified the players in the dataset into premium (paying) and non-premium (non-paying) as well as social and non-social players. A social player was someone who sent at least one friend request to another player, and the category of premium includes all players who made at least one purchase. They found that the factor of current absence time is the strongest CLV predictor. Also, players who play more consistent in the first week were more likely to pay. These results suggest that social players are less likely to convert to premium players as they advance through the game by using their social connections. However, they are still creating value by leveraging network effects and thus advertising the game. In general, these results are counterintuitive with past research that indicated a connection between likeliness to purchase and social interactions. The authors noted that this may be the case due to the nature of the game being a casual game with superficial social interaction mechanics. Thus, their findings may not apply to games with more complex social interactions such as Massively Multiplayer Online Role-Playing Games (MMORPG) [8].

Moreover, Sifa et al. (2015) implemented a model for purchase prediction in a mobile game [30]. They used binary classification to identify players as premium or non-spending users and apply machine learning algorithms on an existing dataset. This method is similar to Hadiji et al. (2014) and Runge et al. (2014). They found that the number of purchases in the past and the amount spent on them are the strongest indicators for future purchases [10, 26]. Other, less significant factors are *(i)* number of in-game interactions, *(ii)* activity related features and *(iii)* total playtime. Furthermore, the authors tried to predict the number of future purchases, to allow more precise overall CLV prediction. Hereby, they found that the amount spent is the strongest predictor for future purchase amount. If players spend relatively large amounts early on, most likely they will continue to do so. Additionally, regional clustering showed that the probability to purchase and the amount of money spent differs between countries. The authors noted that the insights from their study can be used by professionals to strengthen their revenue streams by implementing efficient customer relationship management (CRM) systems. Furthermore, design implications were drawn from the study. For example, it was argued that games should be designed in a way that enforces intense interaction and optimizes for total playtime instead of playtime per session. This indicates, that the overall play experience is more important than just one or two sessions that are perceived as especially positive [30].

3   Conclusion

Game analytics are implemented in novel, powerful tools. They enable companies in the gaming business to predict two of the most important customer-related key data in a reliable manner: player churn and CLV. As a consequence they support strategic design decisions and allow precise and fine-grained player segmentation, which in turn represents a valuable basis for targeting players with appropriate marketing incentives. Together with game analytical metrics, such as retention and acquisition rates, they serve as a common ground for the development of innovative products and of competitive strategies in the gaming industry.

References

1. Aleem, S., Capretz, L.F., Ahmed, F.: Empirical investigation of key business factors for digital game performance. Entertainment Comput. **13**, 25–36 (2016). https://doi.org/10.1016/j.entcom.2015.09.001
2. Brasseur, T.-M., Strauss, C., Mladenow, A.: Business model innovation to support smart manufacturing. In: American Conference on Information Systems 2017, Workshop on Smart Manufacturing, Proceedigs. Association of Information Systems AIS (2017). https://aisel.aisnet.org/cgi/viewcontent.cgi?article=1008&context=sigbd2017. Accessed 27 Mar 2019
3. Clairfield International: Gaming Industry - Facts, Figures and Trends (2018). http://www.clairfield.com/wp-content/uploads/2017/02/Gaming-Industry-and-Market-Report-2018.01-2.pdf. Accessed 27 Mar 2019
4. Davidovici-Nora, M.: Innovation in business models in the video game industry: Free-To-Play or the gaming experience as a service. Comput. Games J. **2**(3), 22–51 (2013). https://doi.org/10.1007/BF03392349
5. Davidovici-Nora, M.: Paid and free digital business models innovations in the video game industry. Digiworld Econ. J. (94), 83–102 (2014). https://ssrn.com/abstract=2534022. Accessed 27 Mar 2019

6. De Prato, G., Feijóo, C., Simon, J.-P.: Innovations in the video game industry: changing global markets. Digiworld Econ. J. (94), 17–38 (2014). https://ssrn.com/abstract=2533973 Accessed 27 Mar 2019
7. Demediuk, S., et al.: Player retention in league of legends. In: Abramson, D. (ed.) Proceedings of the Australasian Computer Science Week Multiconference on - ACSW 2018, pp. 1–9. ACM Press, New York (2018)
8. Drachen, A., et al.: To be or not to be…social. In: Abramson, D. (ed.) Proceedings of the Australasian Computer Science Week Multiconference on - ACSW 2018, pp. 1–10. ACM Press, New York (2018)
9. Drachen, A., Thurau, C., Togelius, J., Yannakakis, G.N., Bauckhage, C.: Game data mining. In: Seif El-Nasr, M., Drachen, A., Canossa, A. (eds.) Game analytics, pp. 205–253. Springer, London (2013). https://doi.org/10.1007/978-1-4471-4769-5_12
10. Hadiji, F., Sifa, R., Drachen, A., Thurau, C., Kersting, K., Bauckhage, C.: Predicting player churn in the wild. In: 2014 IEEE Conference on Computational Intelligence and Games (CIG), pp. 1–8 (2014)
11. Hamari, J., Lehdonvirta, V.: Game design as marketing: how game mechanics create demand for virtual goods. Int. J. Bus. Sci. Appl. Manage. 5(1), 15–29 (2010). http://www. business-and-management.org/library/2010/5_1–14-29-Hamari%2CLehdonvirta.pdf. Accessed 27 Mar 2019
12. Hanner, N., Zarnekow, R.: Purchasing behavior in free to play games: concepts and empirical validation. In: 2015 48th Hawaii International Conference on System Sciences, pp. 3326–3335. IEEE (2015). https://doi.org/10.1109/HICSS.2015.401
13. Flunger, R., Mladenow, A., Strauss, C.: The free-to-play business model. In: Proceedings of the 19th International Conference on Information Integration and Web-based Applications & Services, pp. 373–379. ACM (2017)
14. Komorowski, M., Delaere, S.: Online media business models: lessons from the video game sector. Westminster Pap. Commun. Cult. 11(1), 103–123 (2016)
15. Koskenvoima, A., Mäntymäki, M.: Why do small and medium-size freemium game developers use game analytics? In: IFIP International Federation for Information Processing 2015, pp. 326–337 (2015). https://doi.org/10.1007/978-3-319-25013-7
16. Lee, S.-K., Hong, S.-J., Yang, S.-I., Lee, H.: Predicting churn in mobile free-to-play games. In: International Conference on ICT Convergence 2016, pp. 1046–1048 (2016). http:// ieeexplore.ieee.org/servlet/opac?punumber=7750938. Accessed 27 Mar 2019
17. Macchiarella, P.: Trends in digital gaming: free-to-play, social, and mobile games, Dallas, Texas. Retrieved from Parks Associates (2012) http://www.parksassociates.com/bento/shop/ whitepapers/files/Parks%20Assoc%20Trends%20in%20Digital%20Gaming%20White% 20Paper.pdf. Accessed 27 Mar 2019
18. Marchand, A., Hennig-Thurau, T.: Value creation in the video game industry: industry economics, consumer benefits, and research opportunities. J. Interact. Mark. 27(3), 141–157 (2013). https://doi.org/10.1016/j.intmar.2013.05.001
19. Miller, P.: GDC 2012: How Valve made Team Fortress 2 free-to-play (2012). http:// gamasutra.com/view/news/164922/GDC_2012_How_Valve_made_Team_Fortress_2_ freetoplay.php. Accessed 27 Mar 2019
20. Milošević, M., Živić, N., Andjelković, I.: Early churn prediction with personalized targeting in mobile social games. Expert Syst. Appl. 83, 326–332 (2017)
21. Mladenow, A., Novak, N.M., Strauss, C.: Internet of things integration in supply chains – an austrian business case of a collaborative closed-loop implementation. In: Tjoa, A.M., Xu, L. D., Raffai, M., Novak, N.M. (eds.) Research and Practical Issues of Enterprise Information Systems. Lecture Notes in Business Information Processing, vol. 268, pp. 166–176. Springer, Cham (2016). https://doi.org/10.1007/978-3-319-49944-4_13

22. Oh, G., Ryu, T.: Game design on item-selling based payment model in korean online games. In: Situated Play, Proceedings of DiGRA 2007 Conference, pp. 650–657. The University of Tokyo, Tokyo (2007). http://www.digra.org/wp-content/uploads/digital-library/07312. 20080.pdf. Accessed 27 Mar 2019

23. Park, B.-W., Lee, K.C.: Exploring the value of purchasing online game items. Comput. Hum. Behav. **27**(6), 2178–2185 (2011). https://doi.org/10.1016/j.chb.2011.06.013

24. Perianez, A., Saas, A., Guitart, A., Magne, C.: Churn prediction in mobile social games: towards a complete assessment using survival ensembles. In: 2016 IEEE International Conference on Data Science and Advanced Analytics (DSAA), pp. 564–573. IEEE (2016). https://doi.org/10.1109/DSAA.2016.84

25. Roquilly, C.: Control over virtual worlds by game companies: issues and recommendations. MIS Q. **35**(3), 653–671 (2011). http://www.jstor.org/stable/23042802. Accessed 27 Mar 2019

26. Runge, J., Gao, P., Garcin, F., Faltings, B.: Churn prediction for high-value players in casual social games. In: 2014 IEEE Conference on Computational Intelligence and Games (CIG), pp. 1–8 (2014). https://doi.org/10.1109/CIG.2014.6932875

27. Sandqvist, U.: The games they are a changin': new business models and transformation within the video game industry. Humanit. Soc. Sci. Latvia **23**(2), 4–20 (2015). https://www. lu.lv/fileadmin/user_upload/lu_portal/apgads/PDF/Hum_Soc_2015_2_.pdf. Accessed 27 Mar 2019

28. Seidl, A., Caulkins, J.P., Hartl, R.F., Kort, P.M.: Serious strategy for the makers of fun: analyzing the option to switch from pay-to-play to free-to-play in a two-stage optimal control model with quadratic costs. Eur. J. Oper. Res. **267**(2), 700–715 (2018). https://doi.org/10. 1016/j.ejor.2017.11.071

29. Seufert, E.B.: The freemium business model. In: Freemium Economics, pp. 1–27. Elsevier (2014). https://doi.org/10.1016/B978-0-12-416690-5.00001-4

30. Sifa, R., Hadiji, F., Runge, J., Drachen, A., Kersting, K., Bauckhage, C.: Predicting Purchase decisions in mobile free-to-play games. In: The Eleventh AAAI Conference on Artificial Intelligence and Interactive Digital Entertainment (AIIDE-2015), pp. 79–85 (2015)

31. Urikova, O., Ivanochko, I., Kryvinska, N., Zinterhof, P., Strauss, C.: Managing complex business services in heterogeneous ebusiness ecosystems – aspect-based research assessment. In: The 3rd International Conference on Ambient Systems, Networks and Technologies (ANT-2012), 27–29 August 2012, Ontario, Canada, Procedia Computer Science, vol. 10, pp. 128–135 (2012). ISSN 1877-0509

32. Voigt, S., Hinz, O.: Making digital freemium business models a success: predicting customers' lifetime value via initial purchase information. Bus. Inf. Syst. Eng. **58**(2), 107–118 (2016). https://doi.org/10.1007/s12599-015-0395-z. Accessed 27 Mar 2019

A Decentralized File Sharing Framework for Sensitive Data

Onur Demir$^{(\boxtimes)}$ and Berkay Kocak

Computer Engineering Department, Yeditepe University,
34755 Atasehir, Istanbul, Turkey
odemir@cse.yeditepe.edu.tr, berkay.kocak@std.yeditepe.edu.tr

Abstract. Access management of files which contain sensitive personal data such as electronic health records, financial history of individuals or legal documents is a challenge, because of the variety of participating entities such as legal institutions and privacy of the files. Trust among the entities are generally provided by a central authority such as a government agency. Blockchain technology and distributed file systems provide an opportunity to address this issue without having a central authority to rely on. The lack of a central authority provides a better security as no single entity can control the whole system. However, it brings up other issues. One of the main issues is to manage access rights of sensitive data in the system. In this work, we present a framework for sharing critical document using IPFS and Blockchain technology. Rather than providing framework specific to a single domain such as electronic health records or legal document sharing, we propose a general framework which can be tailored for the specific domains.

Keywords: Decentralized applications · eHealth · IPFS · Blockchain · File systems

1 Introduction

Privacy is an issue in centralized systems in terms of security and trust. Storing files about personal information such as health records, legal documents, or financial history on file systems which can be controlled by corporations or governments is a concern. The incidents of data leaks [1] prove that either extra measures on confidentiality of private data has to be taken or a decentralized solution should be employed. Emerging technologies such as Blockchain and decentralized applications such as IPFs provide alternatives for a better privacy. Even though Blockchain technology provide a decentralized solution for transactions, storing files within the blocks is not a viable solution. Electronic medical records, personal financial history and personal legal documents are typical examples of sensitive data which require frequent updates and sharing among different entities.

There are a number of concerns about how to perform the updates and sharing of sensitive data. The documents has to reflect the full history of their corresponding owner. For example, when a institution requires to access financial

© Springer Nature Switzerland AG 2019
M. Younas et al. (Eds.): Innovate-Data 2019, CCIS 1054, pp. 142–149, 2019.
https://doi.org/10.1007/978-3-030-27355-2_11

history of a person, all documents about the person has to be presented without any exception. Therefore, integrity of personal data has to be present. The files are private, confidentiality of the data has to be protected by encryption. They should also be kept in a structure so that all files has to be linked to each other in chronological order. The access to files has to be logged. The entities who had accessed the files can be verified by screening the logs when needed.

The aim of this work is to provide a decentralized framework that allow sharing of sensitive information using Blockchain technology and a distributed file system.

1.1   Blockchain

Blockchain technology has been introduced by an author using the alias of Nakamoto [2] and lead to a series of developments that shaped decentralized applications later on. Blockchain is a decentralized technology which is built around a data structure that provides a verifiable, immutable, distributed ledger mechanism. Bitcoin and similar cryptocurrencies are regarded as the first generation of Blockchain technology.

The second generation of Blockchain technology enabled users to build decentralized applications (dApps) using Smart Contracts running on a decentralized virtual machine (Ethereum VM) [3]. dApps are verifiable, autonomous, secure and stable applications. They lead to development of applications that does not need a third party to establish a trust mechanism between the users of the application. The original ideology of Blockchain technology envisioned a public framework which will eliminate a central authority [4]. However, there are some hybrid approaches also evolved along with dApps where Blockchain technology can be used for private or semi-private networks. These approaches may require more privileged nodes in the network that regulate the end nodes and their transactions. The original framework is referred as permisionless and the emerging approach is referred as permissioned approaches. Ethereum platform provides a mechanism to construct trust between parties without using a central authority. This mechanism, which is called Smart Contracts, can be programmed by high level languages which can run on distributed nodes and the output can be stored in Ethereum blockchain. For running Smart Contract the parties in the transaction should pay the cost of the transaction by Ethereum platform's virtual currency. The fee covers both computational and storage costs. Therefore, if you need to store large amounts of data, Smart Contracts will not be a cost effective solution. In practice, dApp developers use third party solutions for storage.

1.2   IPFS

Interplanetary File System (IPFS) is a distributed file system which uses content-addressable naming convention. The contents of a file are hashed that are used to address them universally. The files can not be modified once they are created. IPFS does not have an access management layer since it serves to the public domain. A file can be accessed by anybody that know its content name. IPFS provides an alternative for public services such as HTTP and FTP [5].

Blockchain frameworks such as Ethereum has limited capacity to store data within their distributed ledgers. When a block is added to the ledger it will not be deleted forever. Therefore for decentralized application that require storing large amounts of data external solutions must be used. As the decentralized applications should not depend on centrally control systems, using centralized servers or privately owned distributed solutions such as cloud servers will compromise the privacy of the application. For this reason IPFS and similar systems are the proper solutions for storing large amounts of data.

Access control of IPFS has not been implemented natively. Therefore applications that are going to use IPFS generally built custom mechanisms for access control or use it as it is, meaning that their files are public. There are a number of attempts to build frameworks over IPFS and Blockchain such as [6,8,13] IPFS is also suggested to be used for storing transactions in a Blockchain as in [10]. The primary difference with our framework with earlier work is that our framework is designed with the needs of private data which are explained in the following section.

2   Related Work

There are several proposed solution for decentralized access control using Blockchain. There is a survey of decentralized access mechanisms in [9] for large-scale environments. In [7] and [11], data sharing mechanisms are proposed for e-Health systems using Blockchain. In [12], a solution for access control using IPFS is provided similar to our work. In this section we will explain the extra features of our framework.

Our framework is composed of a middle layer for file sharing over IPFS and Ethereum VM. Our main motivation for this framework is to provide a decentralized solution for file sharing between entities where the access to data is controlled, logged and verifiable. We propose a system where securely storing sensitive files and management of them can be done with the following features:

Access Control: The files has to be accessed by only the identities that have the permission. The files can only be read once they are created. The rights for accessing files are created and maintained by Smart Contracts.

Logging: Access to files has to be monitored and logged. Moreover, because each access has to be approved by data owner, the consent of releasing sensitive information is also logged. This feature is essential especially for electronic medical records.

Complete History of an Identity: The files will constitute a history record for an individual. Any missing record or file can hide information about someones history, so it is essential to manage files, that belongs to the same timeline of an individual, as a single entity. For example if a person wants to share credit history files with another institution, all files that are belonging to the same timeline will be shared as a group.

3   A File Sharing Framework for Sensitive Data

In this section we will present the general design of our framework.

There is no personal identity in Ethereum network. All entities in our framework are represented by their digital signatures their wallets. The wallets in a Blockchain platform provides account information and history of all transactions for a given individual. However the wallet addresses and real personalities are not explicitly tied into each other. Therefore even tough all sharing history of an individual are recorded on Blockchain, it is not possible to reveal the real identity of a person by third parties.

Our framework use Metamask browser extension [14] as the wallet. The users of the platform can login into their wallets and start using the framework over browsers. As seen in Fig. 1, our framework acts as a layer over IPFS. Even though the files are stored in IPFS, users cannot access them directly. We provide a browser application as an interface to the framework. Users can see their personal files, permission logs and access logs in their browser. The transaction regarding the access permissions and access logs are stored in Ethereum blockchain.

The access rights are handled as transactions. There are three user roles during transactions: *data owners*, *data creators*, and *data requesters*. Users of the framework can change their roles based on the transactions they are in. Data owners are the user type that determine the access rights of the files. Our model currently does not support co-ownership. However there might be cases where a medical test contains sensitive information for more than one person such as DNA compatibility test for marriage. We plan to address co-ownership in the future, where access rights are managed by more than one data owner. The second user role is data creator who has the right to create files that contains sensitive information. This role represent medical, financial or legal institutions where personal data are generated. The data owners can also act as data creators as well. For example, by filling up a questionnaire, a data owner can create sensitive data which is not necessarily require involvement of an institution. Data creators should interact with data owners to create a file. They do not have any right after the creation of the file. The final user role is data requester. Access to files are given to these users by the data owner. There is only one type of permission for data requester: read. The permission can be given for a specific period of time or it can be revoked by the data owner later on.

Access rights can be given by only data owner. There are basically three types of permissions: *read, revoke* and *create*. When a data creator wants to create a file, create permission is requested from the data owner. Since the files are stored in a public server, when a file is created it is encrypted by a secret key. The key generation, encryption and file upload operations are handled by our browser application. The secret keys are stored in meta-data file which is explained in Sect. 3.1 Once a file has been created, it will never be deleted or its contents are modified. The files of a data owner constitutes a history record of an individual. Hence it is essential to relate all of the files owned by an individual. Whole history of a user is also kept in meta-data file. When an access right is given to a data requester, all files of the data owner can be read by the requester.

Partial sharing of a user's history is also possible based on the creation time of the files. However in that case, the data requester is notified that the data does not cover the full history. The current implementation of the framework does not support partial sharing yet.

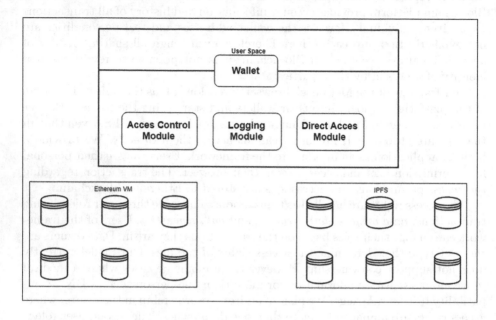

Fig. 1. The architectural layout of the framework.

Figure 1 shows the architectural components of the system. The users are identified by their private keys in the framework. There are three modules that takes care of access rights, logging and direct accesses. When a new file is added into system or when a right is requested, users interacts Access Control Module. Logging Module is used to report logs of the file system. Logging Module gets the log date from a cached copy of transaction records in Blockchain. Direct Access Module is used when a data requester tries to read a file. It is also accessed through the browser application.

All permission requests and grants are recorded in Blockchain using Smart Contracts. For an individual it is possible to keep track of the users who had accessed his/her files in the framework. This will provide a full traceability of access history of users and enable a transparent log of accesses. The permission requests and grants are handled by smart contracts and recorded in Blockchain. The request may have a timeout period so the permission can be revoked automatically. If there is no timeout for the grant, the permission can be revoked by the data owner by another Smart Contract. Currently the framework supports all types of files, however there is a size limitation which is 1.4 GB.

3.1 Meta-data File

We use a meta-data file as a data structure to organize all personal files of a user, cryptographic keys, access rights and other file related data for every single user. Meta-data file can be only modified by Smart Contracts. It is created when a user starts using the framework and its hash name in IPFS is stored in Blockchain.

When the user logs in, user's meta-data file is retrieved to browser application. When a data requester is granted a read right, a copy of the files are encrypted by a secret key. When the requester wants to read files, the secret key of the files are revealed through a Smart Contract to the requester. Once a user gets the secret key, user can access the files using browser application. When the right is revoked, the files are encrypted again by a new secret key. As the content-names of files depend on their contents in IPFS, once a file is encrypted with another key, the name of the file changes and becomes unavailable to public. Meta-data file can be partitioned for context of the files, so that legal documents, financial history and eHealth records can stored at the same time without interfering with each other. The number of contexts can be extended by the users. The secret keys of these context will be different from each other.

4 Evaluation

The performance of our framework can be evaluated in two aspects. The first aspect is the speed of transactions such as granting access rights which are performed by Smart Contracts and the second aspect is speed of file operations over IPFS. The first aspect depends on the amount of Ether paid for the transaction and the load of the total network.

We have used a local Blockchain for testing purposes where transactions are handled instantly. However, in the public Ethereum Blockchain, average time for a transaction to commit is around 15 s which can take longer in the range of minutes to hours if the load on the network is high. Therefore, the performance of transactions are slow with respect to conventional database transactions. We also tested the performance of IPFS performance for file uploads as seen in Fig. 2. The upload performance shows linear behaviour with respect to the file size.

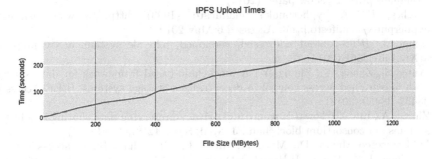

Fig. 2. IPFS upload performance.

5 Future Work

The cumulative data of a number of users can be used for research. In order to solve the privacy issues of personal data, the data can be anonymized by a middle layer service. Even though the current architecture does not consider such a service, we consider providing such a service with the consent of the users. In a situation where an specific type of information is sought in personal files, gathering, organizing and anonymization of the data can be provided as a service.

As noted earlier, we are also planning to provide support for partial access rights for reading a subset of data owner's data. We also considering for a solution for sharing of files with multiple owners.

6 Conclusion

The need for sharing files in a trusted, secure and verifiable environment is necessary where storing and revealing sensitive information about individuals need to be regulated. In our framework, we designed and implemented a data sharing framework over IPFS and Ethereum where transactions are stored in a permissionless Blockchain. Our framework covers the basic needs of electronic medical records, financial and legal histories. We plan to improve our work by providing finer grain features for access rights and anonymity. Our framework supports files of any types up to 1.4 GB of size. The performance of the framework depends on underlying IPFS and Ethereum network performance. The framework is designed for a permissionless Blockchain, however it can be ported to a permissioned Blockchain by having a user authentication layer.

References

1. Greenberg, A.: Hack Brief: Yahoo Breach Hits Half a Billion Users. Wired (2016). https://www.wired.com/2016/09/hack-brief-yahoo-looks-set-confirm-big-old-data-breach/. Accessed May 2019
2. Nakamoto, S.: Bitcoin: a peer-to-peer electronic cash system (2008)
3. Wood, G.: Ethereum: a secure decentralised generalised transaction ledger. Ethereum project yellow paper (2014)
4. Hughes, E.: A cypherpunk's manifesto (1993). http://www.activism.net/cypherpunk/manifesto.html. Accessed 6 May 2019
5. Benet, J.: IPFS-content addressed, versioned, p2p file system. arXiv preprint. arXiv:1407.3561 (2014)
6. Wang, S., Zhang, Y., Zhang, Y.: A blockchain-based framework for data sharing with fine-grained access control in decentralized storage systems. IEEE Access **6**, 38437–38450 (2018)
7. Zhang, A., Lin, X.: Towards secure and privacy-preserving data sharing in e-health systems via consortium blockchain. J. Med. Syst. **42**(8), 140 (2018)
8. Di Francesco Maesa, D., Mori, P., Ricci, L.: Blockchain based access control. In: Chen, L.Y., Reiser, H.P. (eds.) DAIS 2017. LNCS, vol. 10320, pp. 206–220. Springer, Cham (2017). https://doi.org/10.1007/978-3-319-59665-5_15

9. Miltchev, S., Smith, J.M., Prevelakis, V., Keromytis, A., Ioannidis, S.: Decentralized access control in distributed file systems. ACM Comput. Surv. (CSUR) **40**(3), 10 (2008)
10. Zheng, Q., Li, Y., Chen, P., Dong, X.: An innovative IPFS-based storage model for blockchain. In: 2018 IEEE/WIC/ACM International Conference on Web Intelligence (WI), pp. 704–708. IEEE (2018)
11. Dubovitskaya, A., Xu, Z., Ryu, S., Schumacher, M., Wang, F.: Secure and trustable electronic medical records sharing using blockchain. In: AMIA Annual Symposium Proceedings, vol. 2017, p. 650. American Medical Informatics Association (2017)
12. Steichen, M., Fiz Pontiveros, B., Norvill, R., Shbair, W.: Blockchain-based, decentralized access control for IPFS. In: The 2018 IEEE International Conference on Blockchain (Blockchain-2018), pp. 1499–1506. IEEE (2018)
13. Rifi, N., et al.: Towards using blockchain technology for eHealth data access management. In: 2017 Fourth International Conference on Advances in Biomedical Engineering (ICABME). IEEE (2017)
14. Dhillon, V., Metcalf, D., Hooper, M.: Decentralized organizations. In: Blockchain Enabled Applications, pp. 47–66. Apress, Berkeley (2017)

The Information System for the Research in Carotid Atherosclerosis

Jiri Blahuta[✉] and Tomas Soukup

Silesian University in Opava, Bezruc Sq. 13, Opava, Czech Republic
`jiri.blahuta@fpf.slu.cz`, `tsoukup@centrum.cz`

Abstract. This work addresses the possibility of using a new information system and is focused on reproducibility of the Echo-Index between two non-experienced independent observers for atherosclerotic plaques displayed on B-images. The Echo-Index is a numerical value which should characterize the echogenicity grade in selected Region of Interest. In this pilot study, the level of agreement between two independent observers for this index is investigated. The Echo-Index is computed using software image analysis from own developed application designed for digital image analysis of B-images. The results show that the index is well reproducible in time and between two independent observers as well. The study has been performed on a set of 284 B-images. Each image was analyzed two times by one observer and subsequently two times by another observer independently. The Echo-Index is a starting point to create a decision-making expert information system which should help decide about risk assessment in atherosclerosis of carotid artery. The idea of the system is based on differentiation of echogenicity grade according to Echo-Index value. In consequence, the system should classify the plaques into different classes to early risk assessment in carotid atherosclerosis. If the system works properly, many clinical studies can be performed in future; focused on automatic classification of the plaques displayed in B-images to early atherosclerosis diagnosis.

Keywords: Atherosclerotic plaques · B-image · Echo-Index · Echogenicity plaque

1 Introduction

The presented paper is focused on a web-based information system which is developed for image analysis of atherosclerotic plaques in ultrasound B-image. One of the main parts of the system is an algorithm to compute Echo-Index to further B-image analysis for risk evaluation of the atherosclerotic plaques. This system uses this algorithm as a sub-system which is implemented. The main goal of this information system is to collect large sets of B-images with atherosclerotic plaques from radiology centers in the Czech Republic. The reproducibility of the Echo-Index is a key feature to use the system for the further data processing to analyze risk markers of the atherosclerotic plaques in B-image.

© Springer Nature Switzerland AG 2019
M. Younas et al. (Eds.): Innovate-Data 2019, CCIS 1054, pp. 150–161, 2019.
https://doi.org/10.1007/978-3-030-27355-2_12

2   Atherosclerotic Plaque in B-image

This paper is focused on analysis of atherosclerotic plaque in-vitro displayed in B-image from experimental web information system for the B-image data processing and analysis [1], the plaque is displayed as a structure with different heterogeneity and echogenicity, see Fig. 1.

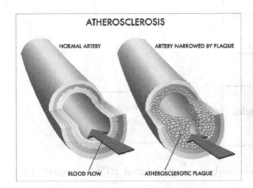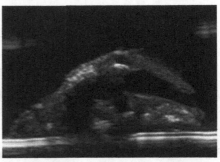

Fig. 1. The principle of atherosclerosis and an example of in-vitro atherosclerotic plaque in B-image (source of the left image: Naturalfoodseries.com)

Simply, atherosclerotic plaque is an abnormal accumulation of material in the inner layer of the wall of an artery. Fat, blood, cholesterol, calcification make common composition of the plaque. Figure 1 shows the example of the plaque in an artery (left) and displayed in-vitro plaque in B-image (right). The progression of the plaque causes arterial stenosis, rupture and finally ischemic stroke. See details in [2].

In general, echogenicity evaluation is one of the most important characteristics of B-image. In general, B-imaging principle is based on display echogenicity according to ultrasound beam reflection [2]. In our application, different echogenicity of the structure is determined by Echo-Index which is the main topic of this research.

3   Echogenicity Grade of Atherosclerotic Plaques

In the case of atherosclerotic plaque, the echogenicity has a special importance. We observe how to measure the echogenicity of the plaque. For this purpose, our developed software [3] is used with extension of Echo-Index which was added later [4,5] for the purpose of using different ROI; not only for substantia nigra. The following Fig. 2 describes the method how to get the Echo-Index.

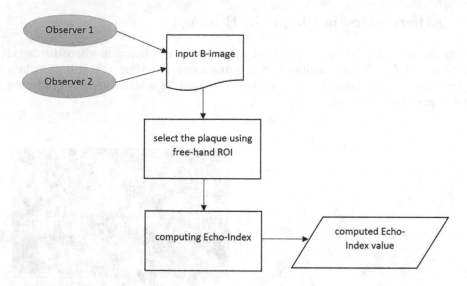

Fig. 2. The Echo-Index from digital image analysis in free-hand ROI is acquired as one numeric value

4   How Echo-Index Is Computed in Selected ROI

The core of the web information system software which has been mentioned was originally developed for investigation of echogenicity grade in substantia nigra structure displayed in B-image. The idea of this software is relatively simple; it is based on computing of decreasing of the area inside a predefined Region of Interest depending on adjustable threshold. With time, within continuous development of the software, the idea of Echo-Index was described. Therefore, one number is obtained on the output, which means the echogenicity grade according to echogenic areas in the B-image. For B-images, on which the echogenicity of the plaque is increased, the Echo-Index should be greater; this simple idea is assumed. In Fig. 1 the example of atherosclerotic plaque is shown. The structure is composed of differently echoic areas. See Fig. 3 on which the plaques with significantly different echogenicity are shown.

4.1   Core of the Algorithm

To understand how the Echo-Index is computed, understanding of the core of the algorithm is needed. However, the algorithm was described in detail in our previous publications, e.g. [3,4]. The basic principle of the algorithm can be described by the following steps:

1. Load the input image (bitmap or DICOM format) which is subsequently converted into 8-bit grayscale depth
2. Select a window in which the examined structure is shown (atherosclerotic plaque)

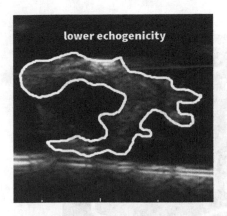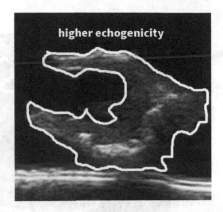

Fig. 3. Lower and higher echogenicity grade of two different plaques

3. Select a Region of Interest (in the case of atherosclerotic plaques, ROI of a free shape), see Sect. 4.2
4. Inside the ROI, the area is computed according to threshold
   (a) The area is computed as the number of remaining pixels after binary thresholding algorithm
   (b) For each threshold T from 0 to 255, the number of pixels is computed
   (c) The number of pixels is converted into real mm^2 according to measuring scale, i.e. the window size in step 2 (the size of $50 \times 50\,mm$ is used)
5. All values of the area for all 256 thresholds are drawn as a "curve" (256 isolated values)

The earlier version of the algorithm ends in this step. See Fig. 4 in which the "curve" is drawn for 2 different images.

4.2 Two Approaches How to Select ROI

To compute the Echo-Index, ROI must be a bordered, closed shape, see Fig. 3. In this application, ROI is a layer above the image.

Manual Drawing the Border. ROI is selected by free-hand drawing with automatic closing as the closed shape. Thus, the ROI is drawn and after a double-click, the shape is closed; the first point is directly connected by a straight line with the last one. Only inside closed region, computing is performed. In case of the plaques in B-image, the auto-closing is used just before the last point. Otherwise, the ROI could not be suitable, see Fig. 4.

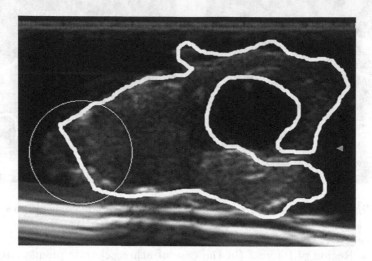

Fig. 4. Missing part of the plaque caused by early automatic closing of the ROI

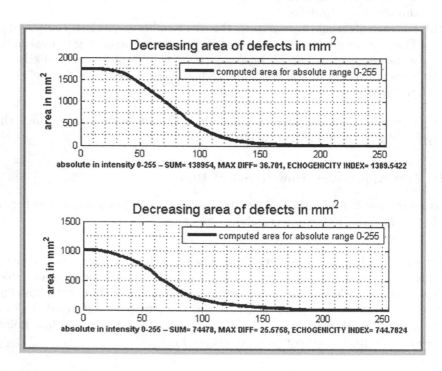

Fig. 5. Decreasing area of remaining area using thresholding and computed Echo-Index for 2 different plaques

Experimental Smart Automatic ROI Selection. Another way to get ROI is to select shape (border) programmatically. New manual drawing of the ROI is an accurate but relatively time-consuming process. For each image, during loading the image to ROI selection, approximately two minutes are lost. Therefore, an experimental solution to automatic ROI selection is under development. This idea is based on black background removal from the image to keep only non-black (echogenic) area. The method is based on the fact, that a solid black area is considered as a useless anechoic area. Simply, only the black area is marked as the background to remove. When the black solid area is removed, the shape of the plaque should be kept as in Fig. 3. The next processing is equal as in the case of manual ROI selection. The idea is shown in Fig. 6. Simply, as in the ROI the remaining foreground is considered and subsequently marked in the input image, see Step 3 in Fig. 6.

Fig. 6. Steps of the automatic ROI selection using background removal

For this purpose, the algorithm called active contour (snake) has been used. It is one of segmentation techniques which is well-applicable for medical images [6]. The principle is based on recognition of background and foreground regions in the image. In addition, number of iterations can be set to achieve better accuracy. However, there are two principal difficulties.

– How large black solid area should be considered as the background?
– How to separate the plaque shape from artery lumen? (Fig. 6, Step 2)
– When the image has increased brightness?

In the case of the first difficulty, the background can be also inside the plaque shape, not only outside. Thus, a minimum area of solid black area should be determined. In the case of the second difficulty, one of the possible solutions is to crop the tube wall from the image. Increased brightness is not difficulty for active contour algorithm. For example, when the background intensity is increased from 0 to 10, this background is still the solid area and the plaque is well-bordered shape as the foreground. This situation is illustrated in Fig. 7. Which implies that the separation of tube wall from the plaque shape is the most important difficulty.

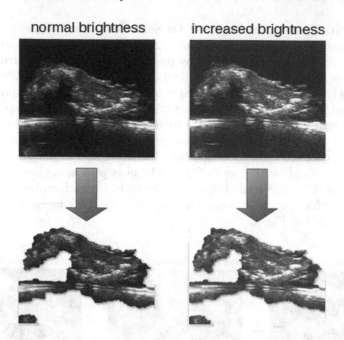

Fig. 7. Increased brightness has no significant influence on active contour to separate foreground

This processing is under development and we must find a way how to select the ROI automatically with required accuracy. If any solution will be found, the process of ROI selection should be reduced to one or two clicks to background removal. In future, black background removal step could be fully automatized. The trouble is sensitivity of the removal; if any small part of the plaque is removed. The pilot study of measurement of the Echo-Index indicates that small differences have no significant influence on Echo-Index value. However, a non-experienced observer cannot draw the plaque border exactly. In essence, this is a similar situation as in the case of an incomplete border caused by automatic closing, see Fig. 4.

4.3   Echo-Index for Free-Hand ROI

In the case of analysis of the substantia nigra, this algorithm was genuinely useful. We can observe the speed of decreasing of the area inside the ROI. For substantia nigra, the ROI was the same for all cases, an elliptical shape with area of $50\,mm^2$. It was sufficient for various clinical studies, e.g. [7] to statistical analysis of the echogenicity grade of the substantia nigra. However, in the case of atherosclerotic plaques, the main difference is the fact that ROI is selected by free-hand shape and the initial area is not equal for all cases, see Fig. 3 above. For this case, the idea of the Echo-Index has been thought as a number which can describe echogenicity grade inside the plaque; inside the free-hand ROI.

Computing of the Echo-Index is genuinely simple. Let we have 256 values of the area for each threshold A_i for $i = 0, 1, 2, \ldots 255$. The sum of the values is calculated as

$$AREASUM = \sum_{i=0}^{255} A_i \tag{1}$$

and the $AREASUM$ value is divided by 100 and this value is called Echo-Index as follows.

$$ECHOGENICITYINDEX = \frac{AREASUM}{100} \tag{2}$$

For example, in Fig. 5, the $SUM = 138954$ (1) and when we divide this value by 100, the $ECHOGENICITYINDEX$ is 1389.54 (2). Due to the principle of binary thresholding, for lower echogenicity grade, the Echo-Index should be lower and for higher echogenicity the Echo-Index should be higher, see Fig. 3 as the example of different echogenicity grade. This is an assumption which proceeds from the principle of binary thresholding. Thus, in the case of low echogenicity, for low T threshold the computed area should be very low and vice versa. In consequence of this principle, the sum for low echogenicity is low and for high echogenic ROI the sum is higher. During the future work it should be confirmed whether this idea is correct or not; especially in comparison with visual assessment.

5 Reproducibility of the Echo-Index Value in a Pilot Study

The main goal of the pilot study [8] was to prove the reproducibility of the index. Within the pilot study, total of 284 B-images were analyzed using this software with this following conditions:

- 2 independent non-experienced observers measured all image two times
- each observer measured for 2 weeks; for one week the first has been performed (284 images) and for following week the second measuring has been performed (the same set of 284 images)
- all images have the same resolution but the algorithm can be used for different resolution

The reproducibility of the Echo-Index has been proved as well-acceptable in general considering inexperience in neurosonology of both observers. The main core of web information system is running on the server and clients have access as users. In addition, the Echo-Index does not evince significant difference in case of the same image analyzed by each observer; each of them draws ROI slightly in different shape. Thus, the index is not significantly sensitive to small changes of ROI. However, some ROI were drawn incorrectly (but similarly) due to inexperience of the observers. Within this study performed by non-experienced observers, the Echo-Index can be considered as a stable feature. Nevertheless, the basic idea was based on "smaller Echo-Index means lower echogenicity" which was not generally confirmed from point of view of an experienced sonographer. Table 1 shows an example of computed Echo-Index for 22 images between 2 observers; there are no significant differences due to similarly drawn ROI. It should be noted that in this study each ROI has been drawn manually not using automatic selection.

Table 1. An example of measured Echo-Index between 2 observers.

IMAGE ID	Echo index observer 1	Echo index observer 2
100027	1229.35	1333.48
105520	1299.29	1371.01
128207	2036.73	1836.71
131283	1386.88	1385.05
137930	962.85	1303.37
144329	977.41	798.63
146994	1180.13	1361.27
153036	774.99	847.06
159463	1464.88	1458.29
160177	1351.90	1318.61
160507	1330.93	1476.99
162265	484.23	493.64
163803	450.52	501.71
197476	747.57	825.111
198052	570.93	653.80
202612	1166.04	1262.82
204891	1213.74	1276.81
214312	1741.40	1817.29
217636	1011.16	1070.25
218762	1142.91	1130.01
220574	893.34	832.47
225903	1280.51	1520.60

5.1   Achieved Results to the Reproducibility Assessment and Using Echo-Index

The pilot study of the web-based information system is focused on computing Echo-Index showed this value is well-reproducible feature between independent observers. In addition, some images from the set were also analyzed by an experienced sonographer; ROI drawn precisely; and very similar values for Echo-index have been achieved.

5.2   Basic Statistical Analysis of Computed Echo-Index Values

To evaluate the reproducibility, the following statistical descriptors were analyzed from 286 B-images:

- range, variance for Echo-Index values measured by observer1 - first measurement R_{11}, var_{11}
- range, variance for Echo-Index values measured by observer1 - second measurement R_{12}, var_{12}
- range, variance for Echo-Index values measured by observer2 - first measurement R_{21}, var_{21}
- range, variance for Echo-Index values measured by observer2 - second measurement R_{22}, var_{22}
- maximum and mean difference between two observers $max(obs)$, $mean(obs)$
- correlation coefficient between Echo-Index values from 2 observers r_{obs}
- maximum and mean difference between measured values from observer1 between two weeks $max(obs1_{2w})$
- maximum and mean difference between measured values from observer2 between two weeks $max(obs2_{2w})$
- correlation coefficient of the Echo-Index values from observer1 between 2 weeks r_{obs1}
- correlation coefficient of the Echo-Index values from observer2 between 2 weeks r_{obs2}

The following results have been obtained, see Table 2.

Table 2. Basic statistical analysis to reproducibility assessment.

Variable	Value(s)
R_{11}, var_{11}	2423.65, 160215.98
R_{12}, var_{12}	2804.90, 164459.53
R_{21}, var_{21}	2215.96, 158765.22
R_{22}, var_{22}	2855.31, 162005.70
$max(obs)$	281.90 (in absolute value)
$mean(obs)$	59.93
r_{obs}	0.947
$max(obs1_{2w})$	312.44 (in absolute value)
$max(obs2_{2w})$	279.61 (in absolute value)
$mean(obs1_{2w})$	64.72
$mean(obs2_{2w})$	70.17
r_{obs1}	0.894
r_{obs2}	0.912

Maximum values are considered in absolute value because the difference of the Echo-Index between observers or measurement can be also negative. In the case of measurement during the first phase (week), for 220 values from 286, i.e. 76.9%, the difference <100 between observers has been achieved. Similarly, in the case of the second phase of the measurement, for 214 values from 286, i.e. 74.8%, the difference <100 between observers has been achieved. Due to results summarized above, the Echo-Index can be considered as well-reproducible value between 2 independent, non-experienced observers and also between 2 measurements from the same observer.

5.3 Role of the Echo-Index in Future Clinical Studies

This finding is a starting point for the clinical study based on comparison of the reliability of image analysis between visual assessment and computer analysis. Future work will be focused on using Echo-Index in clinical studies based on digital image analysis of atherosclerotic plaques which should be more accurate and more unbiased than visual assessment. This index could be well-measurable value for investigation of risk of atherosclerotic plaques after ultrasound examination according to its echogenicity grade as well.

6 Conclusions and Future Work

Achieved results of the Echo-Index reproducibility show that this value could be considered as a well-reproducible value. The information system for the B-image data analysis could be used for this analysis in medical practice. The main goal is to provide the system across the radiology departments in the Czech Republic. Experienced sonographers can analyze the data in real-time and can save all results for differential diagnostics. Image data processing of B-images is a key feature of this system. One of main objectives of this system is to create a tool for storing, processing and analysis of the radiological image data. Although the Echo-Index seems like the reproducible value, its value corresponding with real echogenicity grade (determined by the experienced sonographer) must be thoroughly examined. As the second phase, there is an idea to create a decision-making expert system using knowledge base of echogenicity grades determined by experienced sonographer. In the very beginning, three classes of the echogenicity grade could be used; **Low** (no progress, anechogenic plaque), **Medium** (beginning of the progress, to examine the plaque) and **High** (advanced atherosclerosis, high risk of hemorrhage resulting in ischemic stroke). For example, according to measured values in Table 1, the classes could be determined using the following IF-THEN-ELSE rules.

```
if (echoindex < 500)
    echo_class = "low";
elseif (501 < echoindex <= 1500)
    echo_class = "medium";
else echo_class = "high";
```

This rule-based system could be implemented as a modular subsystem. This is the second phase to create a system using decision-making algorithm to risk assessment of the plaques based on reliable image analysis.

Acknowledgments. The study was supported by the project LQ1602 IT4 Innovations in science and by the project MSK RESTART nr. CZ.02.2.69/0.0/0.0/18-058/0010238. The input data for processing was supported by Ministry of Health of the Czech Republic by grant nr. 16-30965A and 17-31016A.

References

1. Azar, R.N., Donaldson, C.: Ultrasound Imaging (Radcases), 1st edn., Kindle Edition. Thieme (2014). ASIN: B00SRLKPOU
2. Saijo, Y., van der Steen, A.F.W.: Vascular Ultrasound. Springer, Heidelberg (2012). https://doi.org/10.1007/978-4-431-67871-7. ISBN: 978-4431680031
3. Blahuta, J., Cermak, P., Soukup, T., Vecerek, M.: A reproducible method to transcranial B-MODE ultrasound images analysis based on echogenicity evaluation in selectable ROI. Int. J. Biol. Biomed. Eng. **8**, 98–106 (2014). ISSN: 19984510D
4. Blahuta, J., Soukup, T., Cermak, P.: How to detect and analyze atherosclerotic plaques in B-MODE ultrasound images: a pilot study of reproducibility of computer analysis. In: Dichev, C., Agre, G. (eds.) AIMSA 2016. LNCS (LNAI), vol. 9883, pp. 360–363. Springer, Cham (2016). https://doi.org/10.1007/978-3-319-44748-3_37
5. Blahuta, J., Soukup, T., Martinu, J.: An expert system based on using artificial neural network and region-based image processing to recognition substantia nigra and atherosclerotic plaques in B-images: a prospective study. In: Rojas, I., Joya, G., Catala, A. (eds.) IWANN 2017. LNCS, vol. 10305, pp. 236–245. Springer, Cham (2017). https://doi.org/10.1007/978-3-319-59153-7_21
6. Hemalatha, R.J., Thamizhvani, T.R., Babu, B., Chandrasekaran, R., Dhivya, A.J.A., Joseph, J.E.: Active contour based segmentation techniques for medical image analysis. In: Koprowski, R. (ed.) Medical and Biological Image Analysis. IntechOpen. https://doi.org/10.5772/intechopen.74576
7. Skoloudik, D., et al.: Transcranial sonography of the insula: digitized image analysis of fusion images with magnetic resonance. Ultraschall in der Medizin, Georg Thieme Verlag KG Stuttgart (2016)
8. Blahuta, J., Soukup, T., Skacel, J.: Pilot design of a rule-based system and an artificial neural network to risk evaluation of atherosclerotic plaques in long-range clinical research. In: Kůrková, V., Manolopoulos, Y., Hammer, B., Iliadis, L., Maglogiannis, I. (eds.) ICANN 2018. LNCS, vol. 11140, pp. 90–100. Springer, Cham (2018). https://doi.org/10.1007/978-3-030-01421-6_9

Security and Risk Analysis

Big Data Analytics for Financial Crime Typologies

Kirill Plaksiy ⓘ, Andrey Nikiforov ⓘ,
and Natalia Miloslavskaya(✉) ⓘ

The National Research Nuclear University MEPhI (Moscow Engineering
Physics Institute), 31 Kashirskoye shosse, Moscow, Russia
kirillplaksiy@gmail.com,
andreinikiforov993@gmail.com,
NGMiloslavskaya@mephi.ru

Abstract. The paper proposes a technique to automate the generation of new
criminal cases for money laundering from crime and financing terrorism
(ML/FT), which are based on ML/FT typologies. At the same time, the paper is
focused not on the existing methods, but offer its own implemented on the basis
of creating various versions of case typologies and further filtering them by the
derived criteria. For this purpose, it is supposed to use Big Data tools. The
successful application of the developed technique is shown on examples of the
commission and VAT carousel schemes. To implement and verify this technique
a program was written that successfully passed the test on case graphs built on
ML/FT typologies.

Keywords: Money laundering and financing terrorism · ML/FT · Typology ·
Big Data · Graphs · Criteria · Boundary checks

1 Introduction

The legalization of money by criminals concerns a number of subject areas: credit and
financial, budgetary relations (including taxes), corruption offenses, drugs and arms
trafficking based on statistics collected by the Financial Intelligence Units (FIU) of
Russia for 2017–2018 [1]. In 2018, 96 billion rubles were withdrawn abroad through
the commission of suspicious operations. Another 326 billion rubles were cashed.
According to the statistics of the Central Bank of Russia [2], operations to advance the
import of goods were used most often for the withdrawal of money abroad in 2018.
They accounted for 24% (23 billion rubles) of the total funds raised. Transfers for
transactions with services (22% or 21 billion rubles) were also popular.

At present, the bank assets are not only cash but also depositors' personal data,
accounts status information and internal organizational information. If the financial
security (FS) is violated, a financial investigation (FI) is summoned for collecting
materials, clarifying information and evidence on the facilities involved in the incident
and establishing a most possible complete picture of what happened. Financial crimes
or, more precisely, the results of their analysis and cataloging can be attributed to Big
Bata because they are based on three main aspects such as volume (a large amount of

© Springer Nature Switzerland AG 2019
M. Younas et al. (Eds.): Innovate-Data 2019, CCIS 1054, pp. 165–178, 2019.
https://doi.org/10.1007/978-3-030-27355-2_13

data), speed (information processing with high speed), and diversity (data diversity and unstructured). When considering money laundering and terrorist financing (ML/FT) crime schemes, all the participants will have their own identification data, various logical interaction connections, lots of money transactions, etc. There are also modified and combined crime schemes. The Financial Action Task Force on Anti-Money Laundering (FATF) and all FATF-type regional groups carefully investigate the typologies of financial crime. ML typologies are the most common patterns for the legalization of criminal proceeds and FT. Data from various security forces are used to identify typologies in different cases [3]. The definition of typologies in criminal cases has been suggested and analyzed by various FIU [4].

Each typology scheme is unique that creates certain difficulties for identification when it comes to the case to classify real criminal scheme. The task of detecting modified and combined cases for purposes of financial intelligence and law enforcement is one of the priorities today, which requires development and automation of methods and algorithms for searching criminals.

The big data array will be the result of the collaboration of all the services involved for further investigation. FI uses information from transactions, transfer sender/recipient data, account numbers, transfer amounts, etc. Other data such as participants' IP addresses, information about their operating systems and applications, devices used, etc. are used by services, whose task is to ensure information security. The total data array turns out to be beneficial for everyone since it allows to see the relationships that were previously unavailable due to limitations of viewpoints. It is necessary to check the entire volume of generated data for consistency (possibility of implementation in real-world conditions) during the generation of cases on the basis of typologies of financial crimes [2] since the reproduction of financial schemes from typologies and the checking criteria are not reflected in the articles available in the public domain.

Therefore the goal of this paper is to develop the necessary conditions for verifying the generated data for consistency. The research's novelty is to generate new ML/FT criminal cases schemes to multiply them obtaining millions of all possible variants of one case, which then can be processed (analyzed and investigated) by Big Data tools. According to Gartner, augmented analytics and graph analytics used in this research are among Top 10 Data and Analytics Technology Trends for 2019 that have significant disruptive potential over the next 3–5 years due to the need to ask complex questions across complex data [5]. In order to improve the previously proposed mechanisms for the financial crimes detection, it is offered to create all sorts of options for existing ML/FT typologies with many variants (synonyms) and then reduce the number of these schemes leaving only those that are implementable in reality. Finding ML/FT schemes requires new rules that can later be used to automate the search process. The following work has been implemented for this purpose: developing an automation algorithm, choosing a language for writing programs and algorithms at an elected language, testing the developed algorithm in order to determine its compliance to the work goal and Big Data tools choice for the subsequent adaptation of the algorithm to work with a large amount of data.

This paper is organized as follows: Sect. 2 deals with the related works considered. Section 3 presents the criteria for verifying the generated data for consistency in 9 considered financial crimes typologies. Section 4 contains the main outcomes of the proposed work and its evaluation. The last section summarises the proposed work and indicates areas for further research.

2 Related Work

The work on the study of financial crimes and ML/FT typologies, as well as attempts to structure the data and develop algorithms for checking ML/FT have been reviewed. The Federal Financial Monitoring Service of the Russian Federation interacts with the EAG (Eurasian Group on combating ML/FT) in the study of typologies with the FATF as well as other FATF-type regional groups with an eye to the international character and modern scale of ML/FT, which represent a global threat to international security.

One of the EAG main tasks is the analysis of ML/FT typologies [4] taking into account the characteristics of the Eurasian region. Knowledge of the ML/FT sources and methods used to finance terrorism both in the EAG states and in the region is necessary for the development and implementation of effective internal control mechanisms in law enforcement and supervisory activities. The Working Group on Typologies carries out work on priority areas and general research management in the framework of the EAG. The quality of research depends largely on the active participation of the Member States and observers in providing information. Works on financial crimes typologies are conducted on the basis of the schemes analysis for the ML/FT legalization and subsequently averaging them and deriving new typologies. To suppress the actions on ML/FT, the Member States conduct national assessments of ML risks (they are divided into 4 groups: high, elevated, moderate and low) [6], where the following actions are taken:

1. Risk and risk areas' identification;
2. Identification of threats leading to risks;
3. Identification of vulnerabilities exploited for the threats' implementation;
4. Risk assessment;
5. Development of risk treatment measures for each group of risks, for example:
   a. Elimination of nominal legal entities: the introduction of mechanisms to prevent the registration of such companies and the use of false individuals for this;
   b. The duty of authorized banks to refuse to conduct foreign exchange transactions if they contradict the requirements of currency legislation or when the client provides documents that do not meet the established requirements;
   c. Implementation of measures for the search and return of funds withdrawn from the country with the use of non-resident companies through civil litigation and so on.

Economic crimes are committed for the purpose of obtaining benefits in monetary terms. Each fraudulent scheme must have a certain criminal intent to do this.

A few criteria (features) were developed to determine the nature of transactions in the banking sector. Their basis is the same, but each state adapts this series to cover its own, local peculiarities and needs of the financial market [7]. Some signs of ML/FT are presented below:

1. The confused or unusual nature of the transaction, which has no obvious economic meaning or obvious legitimate purpose;
2. The discrepancy of the transaction with the objectives of the organization established by the constituent documents of this organization;
3. Repeated execution of transactions, which nature gives reason to believe that the purpose of their implementation is to evade the mandatory control procedures provided for by the Federal Law;
4. Presence of non-standard or unusually complex instructions on the procedure for making settlements that differ from the usual practice used by the given client (his representative) or from the usual market practice;
5. Client's (his representative) refusal to provide the documents and information requested by the organization that the organization needs to comply with the requirements of the law in the sphere of countering the criminally obtained income legalization (laundering) and terrorism financing;
6. Presence of non-standard or unusually complex schemes (instructions) on the procedure for making settlements that differ from the usual practice used by the given client (his representative) or from the usual market practice.

One should carefully approach the question of whether the meaning, which is criminal in nature, has been preserved in the scheme based on this typology in the process of generating new schemes based on typologies. In accordance with what was said above, it is necessary to calculate the criteria for evaluating typologies for consistency for each typological scheme, i.e. highlight the conditions, under which the commission of financial transactions is criminal in nature and then develop boundary checks of the generated data using the established criteria. Chernorutsky [8] and Nogin [9] help to solve this problem of choice in the decision-making theory.

3 Generated Data Boundary Checks

It is necessary to refer to the decision-making theory to create generated data boundary checks. There are two options for describing choices: a criterial language for describing choices and description of choice in the language of binary relations.

Here the criterial language for describing choices was considered first. The general formulation of the decision-making problem is understood as the problem of choosing from a certain set can be formulated as follows. Let Ω be a set of alternatives. Y is a set of possible consequences (outcomes or results). It is assumed that there is a causal relationship between the choice of some alternative $x_i \in \Omega$ and the occurrence of the corresponding outcome $y_i \in Y$. In addition, it is assumed that there is a mechanism for assessing the quality of such a choice. The quality of the outcome is usually evaluated. It is possible to directly evaluate the quality of the alternative x_i. Then the set of outcomes falls out of consideration, and it is needed to choose the best alternative where the corresponding outcome has the best quality rating. Decision making is demonstrated in Fig. 1.

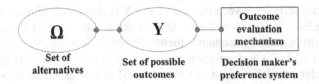

Fig. 1. The task of the decision-making theory

Two options are essential in the decision-making task.

1. It is needed to consider the nature of the connection between the sets Ω and Y.
   (a) Deterministic communication (Fig. 2). There is a unique mapping $\Omega\varphi \to Y$

   $$y = \varphi(x)$$

   $$x \in \Omega, y \in Y$$

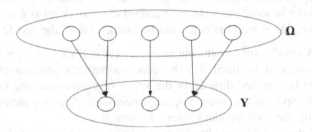

Fig. 2. Deterministic communication

   (b) The relationship has a probabilistic nature (Fig. 3). In this case, the choice of the element x_i guarantees the outcome y_j only with a certain probability P_{ij} where $Sigma_j P_{ij} = 1$, and the decision-making task is called a decision-making task under risk conditions.

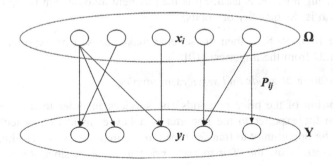

Fig. 3. Probabilistic nature of a communication

(c) If the connection between the sets Ω and Y is non-deterministic and there is no probabilistic information, the decision-making task is called a decision-making task with complete uncertainty then.

2. The second option is a description of choice in binary relations language. Binary relations refer to relations that can be executed or not executed between two elements from the same set. The binary relations language is the second more general criterion language for describing the system of decision maker's preferences. It is assumed that:

1. A separate outcome itself is not evaluated, and criterion functions are not introduced;
2. Each pair of outcomes y_i, y_j can be in one of the following relations.
3. y_i is preferred or strictly dominated by y_j.
4. y_j is preferred or strictly dominated by y_i.
5. y_i is not less preferable (not strictly dominant) than y_j.
6. y_j is not less preferable (not strictly dominant) than y_i.
7. y_i is equivalent to y_j.
8. y_i and y_j are not comparable with each other. It is assumed that the user sets his preferences in some set A. In the standard case, an A-set is a set of outcomes. However, this is with the deterministic connection of the sets Ω and Y.

A = Y or in a multi-criteria outcome assessment A = f(Y), where f = $(f_1, f_2, ..., f_m)$. In the latter case, it is assumed that the decision maker's preferences system (the decision maker) is specified directly in the space of vector outcome estimates. It is assumed that this space is the outcome space if necessary. The user preferences system is specified using the corresponding binary relation R.

The authors derive the logical boundary checks of the data generated on the basis of typologies relying on the decision-making theory.

A. Overestimation of the cost of goods and services in invoices

The key element of this ML/FT scheme is the distortion of the price of goods and services in order to transfer additional amounts from the importer to the exporter [10]. By specifying the cost of goods or services above the "fair market" price in the invoice, the exporter is able to receive funds from the importer since the amount of the payment will be greater than the amount that the importer gains on the open market (Fig. 4).

The following notation is used: Sp is the payment amount, Sm is the market price of the goods, n is the quantity of goods.

Condition 1: if $S_p \gg S_{m*n}$ then the scheme makes sense (the exporter will receive additional funds from the importer) [10].

B. Understating of goods and services in invoices

The distortion of the price of goods and services in order to transfer additional amounts from the exporter to the importer is a key element of this technique. The exporter has the opportunity to transfer funds to the importer by specifying the cost of goods or services below the "fair market" price in invoices since the amount of the

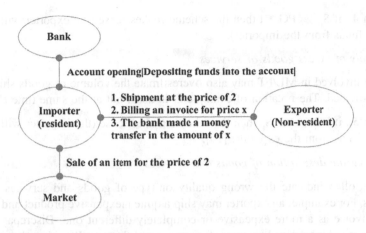

Fig. 4. Overestimation of the cost of goods and services in invoices

payment is less than the amount that the importer gains on the open market. The notation of the example presented above is used.

Condition 2: if $S_{m*n} \gg S_p$ then the scheme makes sense (the exporter will transfer additional funds to the importer).

Thus the "normal" condition for conducting the deals discussed above with no criminal activity observed is the ratio: $S_p \approx S_{m*n}$.

C. Multiple invoicing of goods and services

This technique involves multiple invoices for one international trade transaction. People involved in ML/FT are able to justify multiple payments for the same consignment of goods or the same service by invoicing it several times.

There is the following notation here: i is the unique number of the international transaction, k is the number of invoices issued for the transaction i, b is the number of attracted financial institutions, d is the additional payment transactions under transaction i (payment of the late payment fee and others).

Condition 3: if k > 10 (d > 10) then the scheme makes sense (it is possible to justify multiple payments for the same consignment of goods or the same service).

In addition to Condition 3, one should also check an additional condition to confirm the chosen solution: b > 7 (because of the typology's feature).

D. Undeliverable amount of goods or services

Criminals can underestimate the amount of goods shipped or services provided. Sometimes the exporter does not ship the goods at all, but simply negotiates about the standard clearance of all shipping and customs documents associated with this so-called fictitious delivery with the importer.

There is the following notation here: S_p is the amount of payment indicated in the invoice ($S_p = PG * n$), PG is the price of the goods, n is the quantity of goods indicated in the invoice, t is the quantity of goods shipped after the fact ($n \gg t$).

Condition 4: if $S_p \gg PG * t$ then the scheme makes sense (the exporter will receive additional funds from the importer).

E. Supply of excess goods or services

People involved in ML/FT may also overestimate the volume of goods shipped or services provided. The notation of example D is used, but at the same time $t \gg n$.

Condition 5: if $PG * t \gg S_p$ then the scheme makes sense (the importer will receive additional funds from the exporter) [10].

F. Inaccurate description of goods and services

Rogues often indicate the wrong quality or type of goods and services in their documents. For example, an exporter may ship a quite inexpensive product and include it on an invoice as a more expensive or completely different one. Discrepancies are found between shipping and customs documents and the actually shipped goods as a result. An inaccurate description is also used in the services sale too.

There is the following notation here: PG1 is the price of the actually shipped goods, PG2 is the price of the goods specified in the invoice, n is the quantity of goods, S_p is the amount of payment indicated in the invoice ($S_p = PG2 * n$).

Condition 6: if $S_p \gg PG1 * n$ then the scheme makes sense (the exporter will receive additional funds from the importer).

Condition 7: if $PG1 * n \gg S_p$ then the scheme makes sense (the exporter will transfer additional funds to the importer).

G. Foreign exchange transactions in the black market for the pesos' exchange

The mechanism of the simplest currency operation in this market includes the following steps [2]. The Colombian drug syndicate smuggles drugs into the United States and sells them for cash. The drug syndicate negotiates to a pesos' broker the sale of US dollars below the nominal rate for Colombian pesos. The broker transfers cash in pesos to the drug syndicate from his bank account in Colombia that excludes the further involvement of the drug syndicate in the operation. The peso broker structures or "disperses" the dollar amount in the US banking system in order to avoid reporting requirements and pool these funds on his US bank account. The broker determines a Colombian importer who needs US dollars to buy goods from a US exporter. The peso broker negotiates payment of funds to the US exporter (on behalf of the Colombian importer) from his US bank account. The US exporter ships goods to Colombia. The Colombian importer sells goods (they are often expensive, for example, personal computers, household electronics) for pesos and returns the money to the broker. Thus the broker replenishes the stock of pesos (Fig. 5).

It should be noted that in the course of ML/FT using this approach there are those among defendants who do not intend to commit a crime and who do not suspect that they are involved in a fraudulent scheme.

The drug syndicate is intended to exchange dollars for pesos after the sale of drugs in the United States and getting revenue for them in dollars. It negotiates this with the broker. Cash in dollars is received personally by the broker. The currency in pesos is

Fig. 5. Foreign exchange transactions for the exchange of pesos on the black market

transferred to drug dealers from his account in Colombia. For the services provided the broker charges a fixed percentage of the commission from the amount that was a subject to exchange. This is his economic benefit from the transaction.

There is the following notation here: S$ is cash received from the sale of drugs, SSB is money transferred from a broker's account in Colombia to drug dealers.

Condition 8: the broker will benefit from the transaction: S$ > SSB.

Further, the importer negotiates with the broker so the latter will transfer the dollar amount for the goods to the exporter on behalf of the importer for the purchase of goods. After the exporter's goods are sold, the importer returns the money to the broker including the commission for the services provided by the broker in the total amount.

There is the following notation here: SBE – funds transferred by the broker to the exporter's account on behalf of the importer, SBI – funds transferred from the importer's account to the broker's account.

Condition 9: the broker will benefit from this transaction: SBI > SBE.

In this scheme, the drug syndicate gets an opportunity to legalize those funds, which were received as a result of the sale of narcotic drugs being the main goal for such interaction of the defendants [2], although it does not acquire certain economic benefits in monetary terms.

H. VAT carousel scheme

The carousel scheme (Fig. 6) is widely used at present. A concrete example with concrete figures was examined in order to understand the main principles involved in its operation.

Non-resident company B purchased goods from resident A without VAT for 100 monetary units. Company B must calculate the VAT on the specified transaction for the

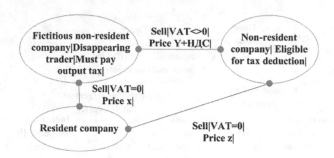

Fig. 6. VAT carousel scheme

purchase of goods, which amount will be simultaneously deductible as VAT input. Company B calculates output VAT at the time of selling goods to non-resident C company at the rate of 18% of 90 monetary units, which gives it 16.2 monetary units with 10 monetary units used to reduce the sale price (the difference between 100 and 90). These 6.2 monetary units are company profits. Then 16.2 monetary units are deducted from non-resident company C as input tax. Non-resident company B pays the output tax to the tax authorities. When company C sells goods back to resident A, VAT is not charged. Company C submits a claim for input tax refund.

The problem for tax authorities is that non-resident company B never pays output VAT. It uses this amount to lower sales prices for non-resident company C. This allows company C to reduce prices for resident company A.

In the indicated deals, the prices are lower than usual. Since non-resident company B does not pay its taxes, the budget loses tax revenues, and ordinary taxpayers suffer as well [10].

There is the following notation here: X is the price of goods when they are sold by resident company A to nonresident B, Y is the price of goods for domestic sale of goods from nonresident B to nonresident C, Z resident company A; Deduction – the amount of tax deduction that a non-resident company will receive; VAT – the amount of value-added tax that non-resident company B must pay to the state budget; RVAT – officially paid value added tax from non-resident company B.

Condition 10: if $(X > Y > Z)$ and (VAT = Deduction provided that RVAT << Deduction or RVAT = 0) then the scheme makes sense (deliberate price reduction to implement the mechanism of the scheme and not full payment of the VAT amount to the budget).

I. Commission scheme

An example of a commission scheme graph is considered in Fig. 7 with the designations: (1) |Supply commission agreement on raw materials, goods | Supply payment|, (2) Supply contract for raw materials and goods, (3) Cash, (4) |Cash | Supply payments for goods, equipment, machinery|, (5) |Supply contract for raw materials, goods | Delivery payment|. A subgraph of this scheme is shown in Fig. 8.

There is the following notation here: Od_i is the i-th company, i is the sequence number of companies, which increases during the resale of goods from one company to

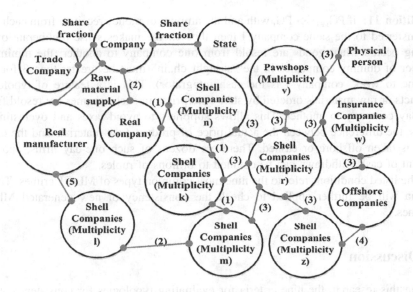

Fig. 7. Commission scheme's graph

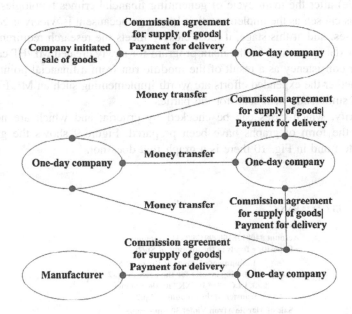

Fig. 8. Commission scheme subgraph

another (initial value: i = 1), PG_i is the total payment amount including the cost of goods and additional expenses sales (for example, commission) to be paid by the i-th company. The quantity of goods n = const.

Condition 11: if $PG_{i+1} \gg PG_i$ with and the amount of money received from each Od_i is transferred to the same company I then the scheme makes sense (deliberate over-pricing occurs when goods are resold from one company to another; the minimum number of dummy companies in the "straight chain" must be greater than 3 for the scheme to work, company I is the cash integrator). This is a feature of typology. Products or raw materials ordered by state enterprises in such schemes are resold from one-day company to another. This allows suppliers to avoid taxes and overestimate prices, the difference between the actual price of products/raw materials and the extra price is taken offshore or cashed. There are dozens of such one-day firms, and the amount of cash withdrawn can be counted to billions of rubles.

The listed conditions refer to the nine most common types of ML/FT crimes. These criteria will be further applied to check the consistency of new generated Ml/FT schemes.

4 Discussion

During this research, the nine criteria for evaluating typologies for consistency were derived. These data boundary checks were implemented as a software module (~ 90 lines of code) after the main cycle of generating financial crimes typologies variants. PL/SQL was chosen as the implementation language because it is widely used to work with databases, and at this stage, it most closely meets the research requirements.

The new data generated by the main program for newly created ML/FT cases were checked for consistency as a result of the module run from a financial point of view, namely, whether the expended efforts are worth implementing such an ML/FT scheme or not, is it sufficiently beneficial for the parties.

For clarity, which cases can be checked by criteria, and which are not, crime patterns in the form of graphs have been prepared. Figure 9 shows the graph that passed the test and in Fig. 10 there is a graph that does not.

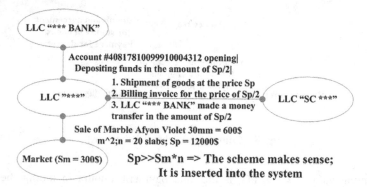

Fig. 9. An example of the boundary check passage. A typology of overstating the cost of goods and services in invoices

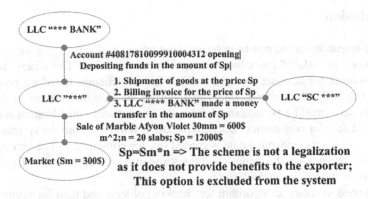

LLC "*** BANK"

Account #40817810099910004312 opening|
Depositing funds in the amount of Sp|

LLC "***"

1. Shipment of goods at the price Sp
2. Billing invoice for the price of Sp
3. LLC "*** BANK" made a money
transfer in the amount of Sp

LLC "SC ***"

Sale of Marble Afyon Violet 30mm = 600$
m^2;n = 20 slabs; Sp = 12000$

Market (Sm = 300$)

Sp=Sm*n => The scheme is not a legalization
as it does not provide benefits to the exporter;
This option is excluded from the system

Fig. 10. An example of unsuccessful boundary check passage. A typology of overstating the cost of goods and services in invoices

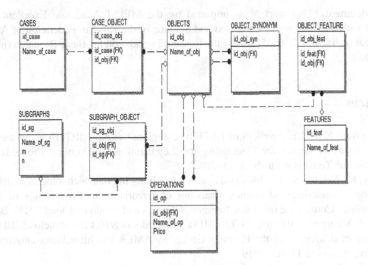

Fig. 11. Relational database

A relational database was created and then used to store the collected data (Fig. 11). The database includes information about the names of cases based on ML/FT typologies using foreign trade operations (CASES), objects involved in these cases (OBJECTS) and their characteristics (FEATURES), subgraphs of the graphs of the main cases (SUBGRAPHS), synonyms of objects (SYNONYMS) and operations performed by the objects of affairs (OPERATIONS).

This database was used to create new cases, which are subsequently verified by meeting the conditions for consistency.

5 Conclusion

During the research, an attempt to improve the efficiency of identifying new criminal schemes based on ML/FT cases built on typologies was made. The criteria for evaluating typologies for consistency were defined for all schemes based on typologies. With the help of these criteria, one can identify the criminal nature of financial transactions performed by persons involved in the cases under consideration and check the generated data for consistency. Unchecked data is removed from the system. Thus a boundary check of the generated data is carried out for consistency. The percentage of verified data depends on the initial conditions specified in the process of generating financial crime typologies variants.

It is planned to create an algorithm for checking objects and their interconnections, the generated data based on financial crimes typologies with the help of the neural network.

Acknowledgement. This work was supported by the MEPhI Academic Excellence Project (agreement with the Ministry of Education and Science of the Russian Federation of August 27, 2013, project no. 02.a03.21.0005) and by the Russian Foundation for Basic Research (project no. 18-07-00088).

References

1. Melkumyan, K.S.: Effectiveness of FATF-type regional groups (RGTF) using the example of the Eurasian Group for Countering the Legalization of Criminal Proceeds and the Financing of Terrorism (in Russian) (2017)
2. Plaksiy, K., Nikiforov, A., Miloslavskaya, N.: Applying big data technologies to detect cases of money laundering and counter financing of terrorism. In: Proceedings of 2018 6th International Conference on Future Internet of Things and Cloud (FiCloud 2018), Barcelona, Spain, 6–8 August 2018, pp. 70–77 (2018). https://doi.org/10.1109/w-ficloud.2018.00017
3. Glossary of definitions of the Eurasian Group on AML/CFT. https://eurasiangroup.org/en/glossary. Accessed 11 Feb 2019
4. EAG Typologies research topics. https://eurasiangroup.org/en/typologies-research-topics. Accessed 11 Feb 2019
5. Gartner Identifies Top 10 Data and Analytics Technology Trends for 2019. https://www.gartner.com/en/newsroom/press-releases/2019-02-18-gartner-identifies-top-10-data-and-analytics-technolo. Accessed 09 Mar 2019
6. National risk assessment for legalization (laundering) of criminal proceeds 2017-2018. Main conclusions. The Federal Financial Monitoring Service of the Russian Federation (2018). http://www.fedsfm.ru. Accessed 11 Feb 2019. (in Russian)
7. Appendix 3 to the Regulation of the Bank of Russia dated December 15, 2014 N 445-P "The requirements for the internal control rules of non-credit financial organizations in order to counter the legalization (laundering) of criminally obtained income and the financing of terrorism" (in Russian) (2014)
8. Chernorutsky, I.G.: Decision-Making Methods. BHW-Petersburg (2005). (in Russian)
9. Noghin, V.D.: Pareto set reducing based on set-point information. Sci. Tech. Inf. Process. **38**(6), 435–439 (2011)
10. "Glitter and poverty" of tax planning during import. http://www.itcon-audit.ru/7.htm. Accessed 11 Feb 2019. (in Russian)

Development of a Model for Identifying High-Risk Operations for AML/CFT Purposes

Pavel Y. Leonov[1](✉) [iD], Viktor P. Suyts[2](✉) [iD],
Oksana S. Kotelyanets[1](✉) [iD], and Nikolai V. Ivanov[1](✉) [iD]

[1] Department of Financial Monitoring, National Research Nuclear University
MEPhI (Moscow Engineering Physics Institute), Moscow, Russia
{pyleonov, OSKotelyanets}@mephi.ru,
ivanov.nikolay.711@gmail.com
[2] Department of Accounting, Analysis and Audit of the Economic Faculty,
Lomonosov Moscow State University, Moscow, Russia
viktor.suyts@gmail.com

Abstract. The article has a high practical significance, which is that the described model of processing and clustering data on banking operations has significantly accelerated the process of identifying suspicious (high risk) among them
and creating a portrait of a client of a credit institution based on its cashless
payments (including his counterparties).

Keywords: Big data analysis · Forensic · Compliance · AML/CFT

1 Introduction

Since credit institutions are the main objects of primary financial monitoring, a large
part of the responsibility for taking measures to counter ML/FT lies with their
employees. In this case, their decisions should be operational and balanced. In this
regard, the development of a model for analyzing the content of payment orders as one
of the tools for performing cashless payments to identify high-risk AML/CFT operations is an urgent and necessary task.

One of the factors that significantly affects the adequacy of the results obtained when
conducting an analysis of any kind is the quality of the data used [1]. In addition to the
final result, the source data has an impact on the speed with which it is received—on the
amount of resources needed to improve their quality ("cleaning"), on the speed of
decision making regarding the choice of analysis methods and software used for this.
Finally, the scope, which reflects the data used, depends on the amount of consequences
that are possible for making decisions based on inadequate analysis results, which are
not carried out on the data that do not reflect the real picture. For example, since
employees of credit organizations have a permission to refuse clients to make a transaction when suspicion arises, making such decisions, which may be erroneous due to
incorrect source data, is fraught with failures for conscientious clients and, conversely,
consent for violators, which, as a result, adversely affect at least the reputation of the
credit institution [2].

© Springer Nature Switzerland AG 2019
M. Younas et al. (Eds.): Innovate-Data 2019, CCIS 1054, pp. 179–192, 2019.
https://doi.org/10.1007/978-3-030-27355-2_14

In this article, some of the details of payment orders, namely the counterparty, date, amount and purpose of payment, are the main data analyzed. Payment orders are filled out and provided by the clients themselves. In this regard, among the reasons for the possible poor quality of data it is worth highlighting, first, the human factor: if the information is entered manually, due to negligence or for other reasons, misprints can be made, information can be entered in the wrong fields or not recorded at all [3]; secondly, the software used: local PC settings are reflected in the representation, for example, dates and numbers with a fractional part (a month or a day comes before, the symbol, "." or ",", separates the integer and fractional part of a number); and, finally, data can be collected by employees from different departments of a client of a credit institution, which only strengthens the first and second possible reasons for the decline in data correctness.

Of course, to achieve 100% data quality, which not only can be filled in manually, but also generally does not have a strictly formalized presentation format, is almost impossible. However, due to the flexibility of the methods used to process them, you can consider some errors (typos) and eliminate their impact on the final result.

2   Primary Data Processing

The first stage of working with data: familiarization with the contents, preparation for import into the database. This stage is implemented mostly in MS Excel. MS Excel is chosen to abstract from any of recent production framework and make accent to details of implemented algorithms and procedures. The task of algorithm programming and creating development production pipeline with related considerations about choosing certain technologies, programming languages, frameworks and performance tuning is not the goal of the article and could be described separately in future.

The initial data were fragments of extracts from bank accounts of a certain organization (hereinafter, the "organization under study"). The organization under investigation has several branches, each of which has accounts in several banks (and branches of these banks). As a result, initially there were 29 .xlsx files, information from which were subsequently merged into one .xlsx file. As a result, this file contains 53044 lines (with headers), reflecting information about the banking operations of the organization under study and its branches in several banks for the period from January 2, 2017 to June 27, 2018.

The data was obtained during the performance of analytical tasks during a financial investigation from the competent authorities. The signed agreement does not make it possible to disclose the source of the data; therefore, the data was impersonal so that, to the maximum extent possible, to ensure work with them without violating legal norms.

Considering the fact of not very large amount of data in the batch authors notice that «big data» volumes could be small but incoming in large volumes [4]. Authors note that the article abstracts from the so-called «Big Data three Vs» and focuses on data analysis algorithms.

Initially, the initial 29 files contained the following fields:

- transaction date,
- name of payer/payee,
- amount of the operation on the debit account,
- amount of the operation on the credit account,
- the purpose of the payment or transfer code.

It is important to note that in banking terms, indicating a certain amount in a debit means debiting funds from the account, and in a loan - crediting. Thus, in each record in one of the amount fields there was a zero, and in the other - the amount of the operation.

The final generated document differs from the original one and contains the following fields (column headings are made in Latin for the convenience of subsequent import to the database):

- organization: in this field, the information about the operation of which branch (or head office) of the organization under study is reflected in this line (i.e. from which of the 29 source files the line);
- bankname: this field is fixed, information about the account operation from which Bank (or branch) is reflected in this line (i.e. from which of the 29 source files);
- oper_date: full compliance with the contents of the field "date of the operation" of the source files;
- counterpart: full compliance with the contents of the field "name of the payer/ recipient of funds" source files;
- oper_type: this field contains only two types of values- "in" or "out", which reflects the transfer (credit) or write-off (debit) of funds (formed artificially through the function MS Excel "IF"—if the original field with the debit account was zero, then the column was written the value "out", otherwise "in"; since then the fields with the debit and credit accounts were removed, to prevent references to non-existent columns, the values were copied and pasted over "as values");
- amt_cur_of_acc: his field contains the value of the amount debited or credited, i.e. full compliance with the contents of field "transaction Amount the debit account, in units of currency of the account or the transaction Amount credited in units of account's currency", respectively (artificially generated through the function of MS Excel "IF"—if in the original box with the debit account was zero, in the column prescribed value of the sum from a column of account credit, otherwise, with the debit account; like the type of operation (oper_type) the obtained values were copied and pasted over the top "as the value");
- dst_text: full compliance with the contents of the "payment Purpose or transfer code" field of the source files.

Figures 1 and 2 show fragments of the final format data file ".xlsx" (organization and bankname fields are hidden).

For a better understanding of how and by correcting what is necessary and possible to improve the quality of the original data, as well as to identify what there are similar

	oper_date	counterpart	oper_type	amt_cur_of_acc
2	27.06.2018	АО "Альфа-Банк"	out	1005,6
3	27.06.2018	Общество с ограниченной ответственностью "Исследуемая организация" Р/С 4070****************	in	5070000
4	27.06.2018	Общество с ограниченной ответственностью "Исследуемая организация" Р/С 4070****************	in	4100000
5	27.06.2018	Ходкевич Светлана Васильевна	out	3777,33
6	27.06.2018	ООО НПФ "Золотая долина"	out	3481549,36
7	27.06.2018	Индивидуальный предприниматель Бугаев Андрей Олегович	out	35100
8	27.06.2018	Общество с ограниченной ответственностью "Исследуемая организация" Р/С 4070****************	in	4930000
9	27.06.2018	ООО "Агама"	out	57750
10	27.06.2018	АО "АЛЬФА-БАНК" г. МОСКВА	out	77349,63
11	27.06.2018	ООО "ИНТЕЛАЙТ"	out	17500
12	27.06.2018	ООО "ПК Дельта"	out	27400
13	27.06.2018	Индивидуальный предприниматель Мустафин Эдуард Альфредович	out	36120
14	27.06.2018	ЗАО "Хэрменс"	out	313410
15	27.06.2018	ООО "ВОТТЛЕР РУС"	out	1323856,74
16	27.06.2018	УФК по Новосибирской области (ФБУ "Новосибирский ЦСМ" л/с 20516Х03090)	out	59427,16
17	27.06.2018	УФК по Новосибирской области (ИФНС РФ по Центральному району г. Новосибирска)	out	835
18	27.06.2018	ООО "ИНТЕЛАЙТ"	out	56151,57
19	27.06.2018	ООО "ИНТЕЛАЙТ"	out	198635,69
20	27.06.2018	Индивидуальный предприниматель Литвинцева Анна Игоревна	out	125000

Fig. 1. Fragment of the final format file «.xlsx» (fields oper_date, counterpart, oper_type, amt_cur_of_acc)

	dst_text
2	(VO80150)Погашение задолж. по комиссии за обслуж. счета в ин.валюте за период с 260518 по 250618 ООО "Исследуемая организация"НДС не обл.
3	Оплата по договору поставки № 3-02/11-02 от 01.11.17Сумма 5070000-00В т.ч. НДС (18%) 773389-83
4	Оплата по договору поставки № 3-02/08-02 от 21.08.17 г.Сумма 4100000-00В т.ч. НДС (18%) 625423-73
5	Отпускные за июнь 2018г.Сумма 3777-33Без налога (НДС)
6	Оплата по с/ф № 625/0000001. 625/0000002 от 25.06.18г.. №626/0000001 от 26.06.18г. за чайный напитокСумма 3481549-36В т.ч. НДС (18%) 531083-80
7	Оплата по договору №1НЛ от 08.05.18 за шкаф-купеСумма 35100-00Без налога (НДС)
8	Оплата по договору поставки № 3-02/10-02 от 27.10.2017г.Сумма 4930000-00В т.ч. НДС (18%) 752033-90
9	Оплата по счету № 406 от 18.06.18г. за стикерСумма 57750-00Без налога (НДС)
10	отпускные за июнь 2018 г сотрудникам ООО "Исследуемая организация" по реестру № 151 от 27.06.2018г согласно приложения № 1 договор № MOSV-531 от 06.04.10 Сумма 77349-63Без налога (НДС)
11	Оплата по сч № 393 от 07.06.18г.- вознаграждениеСумма 17500-00Без налога (НДС)
12	Оплата по сч №99 от 25.06.18 двери. монтажСумма 27400-00В т.ч. НДС (18%) 4179-66
13	Оплата по счету №17 от 08.06.18г за мебель по дог №1004 от 10.04.18гСумма 36120-00Без налога (НДС)
14	Оплата по счету № 43452 от 19.06.18 за коробкиСумма 313410-00В т.ч. НДС (18%) 47808-31
15	Оплата по счету №2 от 26.03.2018 г за шейкерСумма 1323856-74В т.ч. НДС (18%) 201944-25
16	Оплата за проведение исслед-ий продукции по сч №0000-026708 от 22.06.18г. в т.ч. НДС (18%) 9065-16
17	Налог на доходы физ.лиц по ставке в п. 1ст.224 НК. за искл. индив. предприн.
18	Оплата по сч № 442.443.444 от 20.06.18г.- вознаграждение. возмещение расходовСумма 56151-57Без налога (НДС)
19	Оплата по сч № 439.440.441 от 20.06.18г.- вознаграждение. возмещение расходовСумма 198635-69Без налога (НДС)
20	Оплата по сч №28 от 25.06.18г. за разработку дизайн концепции упаковки Сумма 125000-00Без налога (НДС)

Fig. 2. Fragment of the final format file «.xlsx» (field dst_text)

types of operations, the primary analysis was carried out by means of MS Excel. To do this, use the replacement characters (keyboard shortcut "Ctrl+H" indicating the desired character or sequence of characters and replace them) and "Delete duplicates", presented on the "Data" tab.

As can be seen from Figs. 1 and 2, the source data in the counterpart and dst_text fields do not have any uniform formatting: different case, the presence of Cyrillic and Latin, numbers, punctuation marks and other characters; some places are missing and there are typos. At this stage, it is only necessary to estimate approximately what types of operations are contained in the initial data, therefore, surface treatment is sufficient. Information about counterparties does not need this kind of processing, since their supposed possible types depend on whether the person is physical (including individual entrepreneur) or legal (the organizational and legal form of the business entity). Therefore, the purpose of the payment will be considered.

3 Data Grooming

For a start, it is worth sorting the data alphabetically - groups of operations are more clearly visible. After the first iteration, the removal of duplicates from 53043 rows remained 46083 (deleted 7005). Obviously, the information reflected in figures (for example, the amount or date of the operation, the number of the contract or account) is not important for a general understanding of the nature of the operation. After replacing all the digits with "emptiness" and removing duplicates, 13082 lines remained - a quarter of the original data volume. The next group of characters that can and should be neglected is non-alphanumeric characters such as brackets (three types), quotes, apostrophe, percent sign, number sign (also the symbol "N"), forward and backlash, hyphen, dot, comma, etc. (since the characters "*" and "?" are special characters for text masks, you must use the combinations " ~ *" and " ~ ?" to search for them) After this iteration, 12777 lines remain.

Also among the template words and phrases that can be eliminated and not distort the meaning of the content of the purpose of payment at this stage of processing, it is worth noting the names of the months, information about the payment or non-payment of VAT (i.e., combinations like "without (tax) VAT", "including VAT", " not subject to VAT"), the name of the organization under study and its branches, common abbreviations (TIN, KPP, LLC, etc.), some words and abbreviations ("amount", "period", "year (and)", "_" And "g_", "rub", "ul", "kv", "act", etc.), as well as prepositions (through the construction of "predlog"). After this kind of manipulation, 6821 lines remained.

Finally, after deleting words and symbols, sequences of spaces longer than one could remain. They also need to be eliminated: the easiest way is to use the "TRIMS" function (it removes extra spaces not only between words, but also at the beginning and end of a line). After removing duplicates, 4689 lines remain. Thus, 9% of the original text remains.

As a result of the initial data cleansing, it can be concluded that among the groups of operations there are payments (co-payments, prepayments, advance payments) under contracts (for goods, rent, delivery, services, offers, etc.), acts, invoices, invoices and so forth; employee benefits (salary, vacation pay, advances, final settlement upon dismissal); payment of taxes, state fees, penalties, insurance premiums, fines; collection services; commission for banking operations; international settlements; Returns DS. This information will speed up the process of writing regular expressions. It should be noted that the data can be easily adapted to any language, since even if the source data filtering procedures depend on language constructs, they can easily be transferred to any other language due to the simplicity of the procedure itself. In the future in case of using those procedures in production infrastructure they could be upgraded using recent technologies, e.g. full text search engines.

Before importing the data due to some features of the data types supported in the database, it is necessary to convert the numeric variables in the source document (in this case, the amount of the operation) to the format when the "." (Dot) character is used as a separator for the integer and fractional parts. It is also necessary to check whether the data with a dot (part of the sum will be integer) have not been converted into a date format. Then you need to save the file in the .csv format - this is the most suitable format for importing to the database. Since the source file contains Cyrillic characters, the encoding of the created csv file must be changed to UTF-8, otherwise the data may not be recognized when importing.

In most DBMS, data loading via the interface is implemented: the user configures the import, specifying the file location, encoding, field separator, types and, if necessary, the length of variables, the name of the table. Another method used in this case is to create a table for loading data and then load it using an SQL query [5]. The following query structure is used.:

1. deleting the table with the selected name (necessary for quick rewriting);
2. creating a table with the selected name: specifies the field names, type and length of variables (necessary in the case of text data; if the maximum string length is known, there is no sense in increasing the memory allocated for this variable: in this case, the variable oper_type will contain no more three characters);
3. copying data from a csv-file from the designated directory with indication of the field separator and the presence of column headers in the first row;
4. checking the success of the import by uploading data from the created table.
5. Figure 3 shows the code importing the SQL query data, Fig. 4 shows a fragment of the created table (the organization and bankname fields are not shown).

```
drop table if exists payments;

create table payments
(
organization character varying(100) NOT NULL,
bankname character varying(100),
oper_date date,
counterpart character varying(250),
oper_type character varying(3),
amt_cur_of_acc numeric,
dst_text character varying(250)
)
;

copy payments(organization,bankname,oper_date,counterpart,oper_type,amt_cur_of_acc,dst_text)
FROM 'C:\all_data.csv' DELIMITER ';' CSV HEADER;

select * from payments;
```

Fig. 3. SQL query to import data

oper_date date	counterpart character varying(250)	oper_type character varying(3)	amt_cur_of_acc numeric	dst_text character varying(250)
2017-01-09	РОСТОВСКОЕ ОУИ Р/С	in	143000	ПЕРЕЧИСЛЕНИЕ ПЕРЕСЧИТАННОЙ ДЕНЕЖНОЙ НАЛИЧНОСТИ ПРОИНКАССИРОВАННОЙ 05.0 1.2
2017-01-09	ВТБ 24 (ПАО)	in	474557.16	Возмещение ср-в по операциям эквайринга. согласно договору Договор №215 о:
2017-01-09	Индивидуальный пред	out	225100	Оплата по дог возм оказ услуг № 6-03/11-371 от 22.11.2016 за декабрь 2016
2017-01-09	Индивидуальный пред	out	480000	Оплата по дог возм оказ услуг № 6-03/04-236 от 15.04.16 г. за декабрь 201(
2017-01-09	Матюнис Яна Юрьевна	out	48690.25	Оплата по доп соглашению к договору возм оказания услуг № 16-03/04-338 (
2017-01-09	ПАО "ДАЛЬНЕВОСТОЧНЫ	in	111500	Денежные средства по договору инкассация. обработка и зачисление. РКО-16/2
2017-01-09	ИП Лиргута Екатерина	out	480000	Оплата по договору возм оказания услуг № 6-03/02-219 от 01.02.16г. за де:
2017-01-09	ИП Скрыль Варвара Н	out	96000	Оплата по договору возм.оказ.услуг № 16-03/03-110 от 20.03.12г. за декабр：
2017-01-09	ВТБ 24 (ПАО)	in	396424.86	Возмещение ср-в по операциям эквайринга. согласно договору №213 о:
2017-01-09	ИП Гончарова Алена	out	200000	Оплата по договору возм оказания услуг № 6-03/03-215 от 17.03.16 за декабј
2017-01-09	ВТБ 24 (ПАО)	in	402173.14	Возмещение ср-в по операциям эквайринга. согласно договору Договор №212 о：

Fig. 4. Fragment of the source data table imported into the database

Now the data is loaded into the repository and ready for the next stage of processing - writing regular expressions.

The second stage of working with data: the definition of categories of counterparties and payments using regular expressions. Depending on what information is contained in the counterpart and dst_text fields (counterparty and purpose of payment), this or that operation may cause the employee of a credit institution or another analyzing their person to be suspicious of its focus on ML/FT in varying degrees [6]. These degrees were based on suspicion criteria determined by Rosfinmonitoring and expert analysis. However, due to the absence of any clearly defined template for filling in these details of the payment order (there is only a list of information that should be specified, but the method of indication is not set), these data are difficult to understand and quickly identify the essence of the payment. Therefore, it is necessary to automate the process of recognizing the goals of the operation, with what is as accurate as possible.

In short, an algorithm for processing text fields can be described as several iterations of data cleansing through a pattern matching and creating new variables based on them as well as a pattern matching. In this case, both the considered fields are processed simultaneously. As a result, the process of determining the category of counterparty and payment contains four iterations of recognizing the essence of the content of text fields, each of which has several steps, and the final assignment of the degree of suspicion based on these specific categories [8].

As part of the very first iteration of data cleansing, it is necessary to reduce the number of unique counterparties and payment appointments by removing characters, words and phrases that do not bear the semantic load as a whole, without a more detailed study of the contents. Both text fields are subject to processing. Six steps were implemented through nested subqueries: at each subsequent step, the result of cleaning the previous one is processed.

In the first step, all non-alphabetic characters, that is, different from the letters of the Russian and English alphabets, are replaced with a space character. Since the simple deletion of characters can cause the words, initially separated by a space, to merge, each step is replaced by a space, the sequences of which will eventually be eliminated. Also the contents of both clock fields are translated to lower case.

In the second step, the result of the previous step is processed from the counterpart field (so it is more correct to assume that the counterpart1 field is considered, but for convenience, the step numbers in the field names are omitted) replacing with a space eliminates all single characters of both alphabets, t. e. combinations of space-character-space. The month names are eliminated from the dst_text field by replacing with a space: certainly, it is impossible to protect against all possible misprints, therefore the most "explicit" spellings were chosen as templates for matching, which in the process of increasing data cleansing were still complemented with some templates cases of typos (for example, the entry '.e [av] ral.' means that in the first place can be any character, then the letter "e", then one of the letters "a" or "c", then "p", "a" and "l" and any letter at the end - such a record will allow to recognize both "February" and "yyaralya").

In the third step, all single characters of both alphabets are eliminated from the counterpart field, i.e. combinations of space-character-space. The second iteration of such an action is necessary, since in the first, not all the desired combinations will be eliminated. Unfortunately, two iterations of this kind may also not be enough, but most will be recognized, and this will be enough. From the dst_text field, the name of the organization being investigated, and its branches are eliminated.

At the fourth step in the counterpart field, all sequences of spaces from two and above are replaced by one space - this is the final stage of field processing at this step. Regarding the dst_text field, clearing of common words (including prepositions), such as sum, year, act, TIN, period, according to, "g", "from", "for" and others, is framed with a space on the left and space or "line ending" on the right.

At the fifth step, the second iteration of clearing the dst_text field from words of a general nature occurs.

In the sixth step, the dst_text field is cleared of sequences of spaces of two and above in length by replacing with one space.

As a result of the steps taken, 53043 lines left 3128 unique counterparties and 4802 unique payment items.

Second iteration: determination of the category of counterparty and the fact of VAT payment. This iteration has only one step where the CASE selection operator is used and a pattern search is used to determine the counterpart's OPF and whether there was a VAT payment as part of the operation, as well as the formation of a new column of the data table.

Figure 5 shows a fragment of the SQL query to determine the category of counterparty.

```
, case
when counterpart4 ~ '((^| )(ип)( |$))|((инд)(.)*(п[pp][ee]дп[pp]))' then 'ИП'
when counterpart4 ~ '(([pp][oo][cc])*(инк[aa][cc]))|((^| )([oopp][yy]м)( |$))' then 'Инкассация'
when counterpart4 ~ 'б[aa]нк|b[aa]nk|((^| )(вэб)( |$))' then 'Банк'
when counterpart4 ~ '((^| )([yy]фх)( |$))|(([yy]п[pp][aa]вл)(.)*(ф[ee]д[ee][pp])(.)*(к[aa]зн[aa]ч))' then 'УФК (казначейство)'
when counterpart4 ~ '((^| )([yy]фп[cc])( |$))|(почт)' then 'УФПС (почта)'
when counterpart4 ~ '((^| )(фн[cc])( |$))|(н[aa]л[oo]г)' then 'Налоговая'
when counterpart4 ~ '((^| )([oo](3])( |$))|(([oo]бщ[ee][cc])(.)*([oo]г[pp][aa]н)(.)*([oo]тв))|((^| )([oo][cc][oo][oo])( |$))' then 'ООО'
when counterpart4 ~ '((^| )(п[aa][oo])( |$))|((п[yy]бл)(.)*([aa]кци[oo]н)(.)*([oo]бщ[ee][cc]т))' then 'ПАО'
when counterpart4 ~ '((^| )([aa][oo])( |$))|((^[aa]кци[oo]н)(.)*([aa]кци[oo]н)(.)*([oo]бщ[ee][cc]))' then 'АО'
when counterpart4 ~ '((^| )([oo][aa][oo])( |$))|((([oo]тк[pp])(.)*([aa]кци[oo]н)(.)*([oo]бщ[ee][cc])))' then 'ОАО'
when counterpart4 ~ '((^| )(з[aa][oo])( |$))|((з[aa]кр)(.)*([aa]кци[oo]н)(.)*([oo]бщ[ee][cc]))' then 'ЗАО'
when counterpart4 ~ '((^| )([aa]н[oo])( |$))|(([aa]вт[oo]н)(.)*(н[ee]к[oo]мм[ee][pp]ч)(.)*([oo][pp]г))' then 'АНО'
when counterpart4 ~ '((^| )(н[oo][yy])( |$))' then 'НОУ'
else 'Другое' end as agent_type
```

Fig. 5. Fragment of the SQL query to determine the category of counterparty

When searching by a template, the presence of a space or "beginning of line" ("^") on the left and a space or "end of line" ("$") on the right, several variations of matches (for example, for "ip" and for "individual entrepreneur"), as well as taking into account the letters of both alphabets, similar in shape ("a", "e", "o", "p", "s", "y"). The category "Other" is intended for those counterparties whose type could not be determined: in the initial data set, organizations belonging to the English alphabet, as well as individuals, fell into it.

As a result, 33% of counterparties were attributed to banks, 27% to individual entrepreneurs, 19% to LLC; 11% of counterparties are classified as "Other". This completes the processing of the text field with the name of the counterparty.

Figure 6 shows a fragment of the SQL query for determining the fact of payment/non-payment of VAT.

```
, case
when dst_text6 ~ '(ндс нет)|((без)+( )*(налога)*( )*(ндс))|((ндс не)( )*(о)( )*(бл)(агается)*)|((ндс не предус)(.*))|(( б*е*з налога)( нд)*ё)|(без нд*ё)'
    or lower(dst_text) ~ '(ндс)( )*([^0-9])*(0\%)' then 'без НДС'
when dst_text6 ~ '(^ндс)|((в )*(в том числе|том числ е)( )(ндс))|(с ндс)'
    or lower(dst_text) ~ '(ндс)( )*(([1-9]+[0-9]*)*' then 'с НДС'
when dst_text6 not like '% ндс%' then 'нет инфо'
else 'другое' end as nds_info
```

Fig. 6. Fragment of the SQL query for determining the payment/non-payment of VAT

To determine the category of payment in the context of VAT payment, a template search is performed not only by the contents of the column with the last cleaning iteration (dst_text6), but also by the original contents of the payment purpose, because during the minimization of the number of payments falling into the "Other" category, cases were found when the payment or non-payment of VAT was indicated as "VAT 0%" or "VAT 18%" without any adjacent auxiliary words. At the same time, since the first step of the first iteration of cleaning replaces all non-letter characters with spaces, only the abbreviation of VAT remained, according to which it was impossible to draw any conclusions.

As in the definition of the category of counterparty, the "beginning and end of line" symbols were taken into account, framing the gaps and all sorts of variations of writing VAT information.

As a result, 73% of payments were made without paying VAT. One payment fell into the "other" category: it is impossible to foresee all variants of misprints.

In addition to determining the categories of counterparty and the payment of VAT within this iteration, the column dst_text_wtht_nds was obtained, which contains information about the purpose of payment without signs of payment/non-payment of VAT. After such processing, there are 4,621 unique payments.

Third iteration: determination of the category and subcategory of payment. This iteration consists of two steps. In the first step, the payment category is determined (Fig. 7).

```
, case
    when dst_text_wtht_nds ~ '^vo' then 'валютная операция'
    when dst_text_wtht_nds ~ 'займ[и]заем' then 'займ'
    when dst_text_wtht_nds ~ 'эквайрин' then 'услуги эквайринга'
    when dst_text_wtht_nds ~ 'возврат' then 'возврат ДС'
    when dst_text_wtht_nds ~ '(конвертаци|конвс[а-я]* |ком яікон )(.)*(банк|збо|курс|обслуж|расч[а-я]* сч)' then 'комиссия за банк услуги'
    when dst_text_wtht_nds ~ 'комис[а-я]* |ком яікон ' and not dst_text_wtht_nds ~ '^инкас' then 'прочая комиссия'
    when dst_text_wtht_nds ~ '(инкасс)(о)*|(ікм[а-я]* кол (1[а-я]*))|(ваат ([а-я]* экс[а-я]*))|(ваат ([а-я]*) выруч[а-я]*)|( сумк(.)* икк)' then 'инкассационные услуги'
    when dst_text_wtht_nds ~ 'зар[пл](2)ар.*|з( ар)плат[а-я](0,2)|з( )*п( (6)|зарабоги|сотрудни|зольн|отпуск|пособи|сон(ь)*ые' and not dst_text_wtht_nds ~ 'оптлан|ин(.*)осн'
        then 'выплаты сотрудникам'
    when dst_text_wtht_nds ~ '((гос)*(пошлин))|(( (*)налог)|(страх)|(регистр.*стр )|(свид.* гос.* регистр.*)|штраф)' then 'налоги/страховые взносы/пошлины/штрафы'
    when dst_text_wtht_nds ~ '(за)*кол[ем][а-я]*' then 'выплаты по ПОДПУ/задолженности'
    when dst_text_wtht_nds ~ 'возмзр[а-я]*(.)* (хранен|хранит)' then 'возмещение по договору хранения'
    when dst_text_wtht_nds ~ 'возм[а-я]* ' then 'возмещение'
    when dst_text_wtht_nds ~ 'процен[а-я]*(.)* счет' then 'проценты по счету'
    when dst_text_wtht_nds ~ '(пре)*(за)*(о)*плат[а-я]*|услуг[а-я]*|акт|дог[а-я]* воз[а-я]* ок[а-я]* усл[а-я]*|платеж( |*)аванс|( (*)опла по |договор'
        then 'оплата договор/сч-факт/акт'
    else 'другое' end as oper_info
```

Fig. 7. Fragment of the SQL query to determine the category of payment

The categories included currency operations, loans, acquiring services, cash refund, banking services and other commission fees, collection services, employee benefits, taxes, duties and insurance premiums, debt payments, storage contracts, refunds, interest payments. by invoice and, the most extensive, payment under contracts/ invoices, etc. All other payments fall into the "Other" category.

In the second step, the categories "currency transaction", "loan", "taxes/insurance premiums/fees/penalties" are defined by subcategories. Since in the "payment contract/account/act" category the main work on identifying a subcategory occurs at the next stage, the first iteration of eliminating common words inherent in such payments (for example, contract, act, services, paid services, - invoice, etc.). For other categories, subcategories are not defined.

In determining the category and subcategory of payment, of course, various variations of the spelling of words and phrases were taken into account, according to which conclusions are drawn about its essence.

The fourth iteration: the definition of a subcategory for payment under contracts, invoices, acts. This iteration consists of five steps and the use of a CASE select statement at the end.

At the first step, the second iteration of the replacement of common words characteristic of this category occurs, in which the previous iteration determined that it was a contractual payment, etc., for gaps. At the moment, they no longer carry a semantic load: the emphasis is on determining which product or services are paid for, and not just the fact of payment.

In the second and third step, the elimination of words with a length of two characters, framed by spaces or the characters at the beginning and end of a line, occurs - these are mostly prepositions. In the fourth step, a sequence of two or more spaces is replaced by one.

At the fifth step, situations are revealed when, after the previous stages of purification, only single words remain (that is, there is only a "beginning of the line" character before a word, and "end of a line" after it) words that do not carry a semantic load, unless they have any clarifications. For example, advance payment, services, work, prepayment, etc. there are cases when it is possible to determine that this payment consists in receiving or deducting a CP for the business activities of the organization, but it is not clear what it is for. Such payments, obviously, can cause more suspicions than those where specifically indicated, for example, "payment for water and glasses."

Then, using the CASE selection operator, a subcategory is defined within the category "payment contract/account fact/act" (Fig. 8).

```
, case
when oper_info = 'оплата договор/сч-факт/акт' and dop_oper_info6 = '' then 'не установлено'
when oper_info = 'оплата договор/сч-факт/акт' and dop_oper_info6 = 'з(а)*нак[а-я]* тов[а-я]*|тов[а-я]* з(а)*нак[а-я]*|франч[ий]з|роялти|[*не]искл[а-я]*( )*пра[а-я]*'
      then 'товарный знак/франчайзинг/роялти/искл права'
when oper_info = 'оплата договор/сч-факт/акт' and dop_oper_info6 = 'оферт' then 'договор оферты'
when oper_info = 'оплата договор/сч-факт/акт' and dop_oper_info6 = 'трансп[а-я]*|перевоз[а-я]*' then 'транспортные услуги'
when oper_info = 'оплата договор/сч-факт/акт' and dop_oper_info6 = 'аренд' and dop_oper_info6 = 'помещ|недвиж|нежил|фасад|зем[а-я]*( )*участ[а-я]*'
      then 'аренда нежилого помещения/недвижимости/земли'
when oper_info = 'оплата договор/сч-факт/акт' and dop_oper_info6 = 'интернет|помещ|абон|связ[ьь]|телефон' and not dop_oper_info6 = 'право нет'
      then 'интернет/связь'
when oper_info <> 'оплата договор/сч-факт/акт' then dop_oper_info6
else 'другое' end as dop_oper_info
```

Fig. 8. Fragment of the SQL query to determine the subcategory for "payment contract/account fact/act"

There are highlighted payments for trademarks and franchising, offer agreements, rental of real estate, provision of transportation services, payment for communication services and access to the Internet, as well as cases where it does not specifically designate the subject of the payment agreement. The remaining payments are determined by the "Other" subcategory. This completes the process of determining the categories of counterparty and payment.

Using a numeric equivalent for text variables is much more convenient from the point of view of future processing, and since various categories can be assigned the same numerical equivalent, since they are equivalent.

The sign of replenishment or write-off of DS is binary: the operation of replenishment is equivalent to zero, and the operation of write-off - to one.

The categories of counterparties and payments, including the fact of payment/non-payment of VAT, on the basis of expert opinion are assigned the degrees of suspicion shown in Table 1.

Table 1. The degree of suspicion of counterparty categories and payments

Degree of suspicion	Category		
	Counterparty	VAT payment	Payment
1	Individual entrepreneur	Without VAT	Interest-free loan, trademark
0,9	–	No info	Debt/indebtedness
0,85	–	–	Interest-bearing loan, refund
0,8	LLC, ANO	–	Storage agreement
0,7	Private educational institution	–	Loan (rate unknown) offer contract, transport services
0,6	JSC, CJSC, OJSC, PJSC	–	Not established (for payment under the contract)
0,5	Other	Other	Other
0,4	Office of the Federal Postal Service	–	Other (for payment under the contract)
0,3	–	–	Rental of premises/land
0,2	Bank	–	Contribution to the authorized capital, collection
0,1	Encashment	–	Employee benefits, Internet/communication
0	Office of the Federal Treasury, Tax	With VAT	Commissions, acquiring, % on account, taxes, insurance payments

This step is final at this stage of working with data. For the convenience of the subsequent creation of variables that will be the basis for cluster analysis, the table is formed from the results obtained in the database.

As noted before this algorithm implementation is the only described at the moment, all necessary changes or additions could be performed within the framework of the described approach. The task of generalizing this algorithm and creating the possibility of its configuration is not the goal of the article and should be considered separately.

4 Data Clustering

The source data, the processing of which is described earlier, is information about the operations of an actual organization. Firstly, it improves the quality of written regular expressions, since the data were not artificially created, and therefore reflect all those factors that are inherent in the real sets of textual information (typos, lack of uniform formatting and pattern). Secondly, it is possible to assess the applicability of this method of processing the contents of payment orders, since Based on these data, an investigation was conducted to identify counterparties, the activity is suspicious in the area of AML/CFT, however, data were processed manually by several analysts (in Microsoft Excel, by filtering table fields and searching for words).

Consequently, as part of the preparation of this article, the clustering of counterparties is more likely a way to assess the quality of text recognition by written regular expressions, but also one of the options for further data processing for suspicious counterparties, since in any case, simple categorization by OPF and in fact payment (including VAT payment) is not enough.

If there are any other ways to determine the type of payment or the categories of counterparties described above [7], the algorithm of actions during cluster analysis will also remain suitable for identifying the most suspicious of them.

As mentioned before, the data set provided for processing represents real information about the banking operations of a real-life organization. Moreover, information about contractors, whose actions of analysts who worked with the data "manually" were considered suspicious, were transferred to law enforcement officers. Subsequently, a criminal case was opened.

In this case, the approach used in the framework of this article is convenient from the point of view of sufficient universality and saving of time resources (the analysis was carried out about seven times faster). The authors note that the resulting deviation in the results from manual execution is compensated for by the speed of the automated solution. Note also that Big Date solutions often have errors due to the chosen algorithms and technologies in production, thus this deviation is acceptable.

Due to the large number of analyzed counterparties, their number was reduced due to consideration of counterparties with a degree of suspicion of 0.5, while conducting operations in the amount of more than 100 thousand rubles and a degree of suspicion of 0.5. The counterparties that met these conditions were also divided into three groups: individual entrepreneurs, limited liability companics and others (mainly various types of joint-stock companies). For each of the samples, the cluster analysis was carried out using the Ward and k-means methods. As a result, 12 cluster analyzes were carried out (three samples, two methods of forming a set of features, two clustering methods). An example of the results of cluster analysis is presented in Fig. 9.

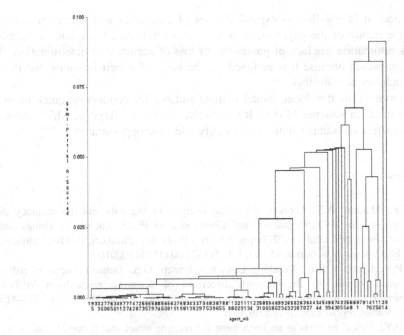

Fig. 9. Clustering dendrogram

The purpose of clustering was to search within a sample of such counterparties that were identified either in a cluster with a small number of objects or were the only objects in the cluster. In this case, the most suspicious were those counterparties that were isolated within the framework of all four cluster analyzes, i.e. on both methods of formation of attribute space and on both methods of clustering.

The results were compared with the results obtained by analysts using other detection methods. A consequence of the effectiveness of cluster analysis was the coincidence of the results obtained with the list of counterparties that were deemed suspicious and questionable during the application of the developed model. It was assumed that the results of the obtained algorithm will be compared with the results of other known algorithms, but, as mentioned above, there are no similar algorithms in the public domain.

5 Conclusion

As areas of improvement is, above all, note the need to improve the source data: they should be more complete and have parameters that uniquely define the counterparty. Also expanding the source data will provide an opportunity to increase model functional: analysis can be carried out not only on the basis of purpose of payment, but also taking into account other characteristics of counterparties; but the introduction of a uniquely defining parameter will allow greater accuracy and more sections to group the data.

In addition, it is possible to expand the set of categories and subcategories to determine the nature of the payment, as well as more automated suspicion determination (full automation has lack of possibility of loss of accuracy of determination of the degree suspicion because it is assigned on the basis of expert opinions that may vary from industry to industry).

At the moment the developed model is most suitable for conducting quick investigations on medium volumes of data. It is intuitive to use and flexible, which allows change parameters for sample limits and quickly add new opportunities.

References

1. Suyts, V.P., Shadrin, A.S., Leonov, P.Y.: The analysis of big data and the accuracy of financial reports. In: 2017 5th International Conference on Future Internet of Things and Cloud Workshops (FiCloudW) 2017, pp. 53–56 (2017) https://doi.org/10.1109/ficloudw. 2017.93, http://gateway.webofknowl…rd; UT=WOS:000435144700010
2. Leonov, P.Y., Leonova, E.K., Veselova, E.A., Kotelyanets, O.S.: General scheme of risk – oriented audit stages. In: Breakthrough directions of scientific research at MEPhI: Development Prospects within the Strategic Academic Units, Knowledge E, pp. 402–415 (2018). https://doi.org/10.18502/keg.v3i6.3021
3. Leonov, P.Y.: Visual analysis as an instrument for creating unreliable financial statements. Sci. Vis. 9(5), 137–143 (2017). Q4. https://www.scopus.com/record/display.uri?eid=2-s2.0-85039864252&origin=resultslist
4. Laney, D.: 3D data management: controlling data volume, velocity and varieety. Application Delivery Strategies (2001). https://blogs.gartner.com/doug-laney/files/2012/01/ad949-3D-Data-Management-Controlling-Data-Volume-Velocity-and-Variety.pdf
5. Leonov, P.Y.: Development of a process for detecting inconsistencies between financial statements and financial transactions (Deals) aimed at identifying signs of money laundering. In: III Network AML/CFT Institute International Scientific and Research Conference "FinTech and RegTech", Knowledge E, pp. 118–131 (2017). https://doi.org/10.18502/kss. v3i2.1533
6. Zhukov, A.N., Leonov, P.Y.: Problems of data collection for the application of the Data Mining methods in analyzing threshold levels of indicators of economic security. In: III Network AML/CFT Institute International Scientific and Research Conference "FinTech and RegTech", Knowledge E, pp. 369–374 (2017). https://doi.org/10.18502/kss.v3i2.1566
7. Shadrin, A.S., Leonov, P.Y.: Risks evaluation of financial-economic activity and their management in the system of economic security of the organization. In: III Network AML/CFT Institute International Scientific and Research Conference "FinTech and RegTech", Knowledge E, pp. 427–435 (2017). https://doi.org/10.18502/kss.v3i2.1573
8. Leonov, P.Y., Kazaryan, S.G.: The usage of analytical SAS tools in the audit practice for risk assessment. In: III Network AML/CFT Institute International Scientific and Research Conference "FinTech and RegTech", Knowledge E, pp. 552–560 (2017). https://doi.org/10. 18502/kss.v3i2.1589

Monitoring System for the Housing and Utility Services Based on the Digital Technologies IIoT, Big Data, Data Mining, Edge and Cloud Computing

Vasiliy S. Kireev ⓘ, Pyotr V. Bochkaryov, Anna I. Guseva⁽⊠⁾ ⓘ,
Igor A. Kuznetsov, and Stanislav A. Filippov

National Research Nuclear University MEPhI (Moscow Engineering
Physics Institute), 31 Kashirskoye shosse, Moscow, Russia
{VSKireev, PVBochkarev, AIGuseva, IAKuznetsov}@mephi.ru,
Stanislav@Philippov.ru

Abstract. This article discusses the development of an information system for monitoring, remote control and support of technical conditions in a housing infrastructure engineering system. The monitoring system is based on the concept of edge computing - cloud computing and is designed to collect and analyze Big data procured via the Industrial Internet of Things (IIoT). Author acknowledges support from the MEPhI Academic Excellence Project (Contract No. 02. a03.21.0005).

Keywords: Big data · Data mining · IIoT · Cloud computing · Edge computing · Information systems for monitoring

1 Introduction

Housing and utility services (HUS) of the Russian Federation is a complex and man - made system that encompasses more than 52 thousand enterprises and employs 4.2 million workers. The share of housing and utility services in the national economy is twenty-six percent; whereas, the wear and tear of municipal infrastructure amounts to more than its' sixty percent, Additionally, seventy-three percent of engineering equipment and 65% of engineering networks are in immediate need of replacement. Thus, the development of an information system that provides monitoring, remote control and technical support to the engineering systems of housing infrastructure is an imperative task. Such information system will assist the engineering system in operational control, maintenance, anticipatory repairs and notification of scheduled and emergency situations.

M. Younas et al. (Eds.): Innovate-Data 2019, CCIS 1054, pp. 193–205, 2019.
https://doi.org/10.1007/978-3-030-27355-2_15

2 Problem Statement

2.1 Housing and Utility Services

Today, the housing and utility services of the Russian Federation are in critical condition. The losses in electricity resources amount to 30–40%; the water resources are depleted by 10–15%; and the gas is down by 60%. There are multiple reasons for such a significant reduction in resources, some are due to illegal connections, some are due to errors in taking down the readings. Many works that research and analyze the existing problems in the sphere of housing and utility services agree that these services have a significant impact on all aspects of society [1, 2]. As a result, monitoring the state of housing and utility facilities is a matter of national security. End-to-end digital technologies such as the Industrial Internet of Things, big data, cloud, fog and boundary calculations, as well as data mining are now the basis for monitoring housing and utility facilities.

2.2 The Industrial Internet of Things

The Industrial Internet of Things (IIoT) is the concept of building information and communication infrastructure, by connecting any non-household equipment, devices, sensors, gauges, automated control system of technological process (ACS TS) to the Internet, and integrating all these elements among themselves. This leads to the emergence of new business models for the production of goods and services and their delivery to consumers. The Industrial Internet is one of the applications of the Internet of Things (IOT), which is a more comprehensive technology that allows both, household and non-household devices to interact via the Internet [3, 4].

The structure of the Industrial Internet technology stack is shown in Fig. 1.

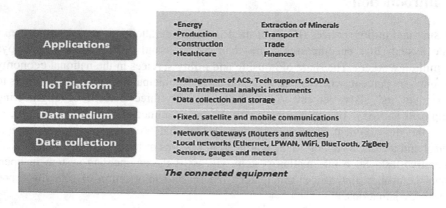

Fig. 1. Technological basis of IIoT

In terms of technology, the Internet of Things includes the following components [5]:

- Devices and sensors able to collect, record, analyze and transmit data over the network;
- Communications means - a network infrastructure that combines heterogeneous communication channels-mobile, satellite, wireless and fixed – the Industrial Internet includes a multitude of wireless options. In various IOT projects, the connection can be based on such technologies as Wi-Fi, Bluetooth, 3G, 2G, LTE, RFID, LPWAN, NFC, ZigBee, LiF, Z-wave, LoRa, etc.;
- IoT platforms from IT vendors and commercial companies designed to manage devices, communications, applications, and analytics. These platforms should contain a development environment and IT security solutions;
- Applications and analytical software-the software layer responsible for analytical data processing, predictive modeling, and intelligent device management;
- Data storage and server systems able to store and process large amounts of information;
- IT services that are based on knowledge of a particular industry and its business specifics and offer solutions in the field of the IoT;
- Security solutions that are responsible for information security of all components of a solution and for the security of the operational process. Since the Industrial Internet implies a close integration of IT and production processes, the task of security goes beyond ensuring smooth operation of the IT infrastructure.

In the context of our work, we are interested in devices and sensors that can capture events, collect, analyze and transmit data over the network, as well as the means of communication that provide two-way connection with monitoring and control devices. We are also very curious about analytical software, responsible for analytical data processing and intelligent device management.

In Russia, the market for IoT technologies is at an initial stage of formation. The total volume of connected devices worldwide is 4.6 billion; Russia accounts for about 0.3% of that amount. The main reason for Russia's lagging behind the pace of world development is that Russian market offers standalone devices and not the complex solutions. There is no integrated end-to-end production; the market lacks the uniform ecosystem [6].

The market development estimates differ significantly from one source to another. According to Gartner analysts, the number of connected devices will reach 21 billion units in 2020, while Intel gives a figure of 200 billion. Despite the substantial differ-ence in estimates, we can assert the high growth rate of the IoT market. This triggers serious interest from commercial companies, large device vendors, platform and application developers, research agencies and national government agencies. Thus, Ovum predicts that in 2019, the total amount of connected devices in the world will reach 530 million units; and, the largest number of such devices will be in the fields of energy and housing, transport, industry, health and trade [5].

A key driver for IoT market growth is predicted to be the continuing decline in the costs of sensors and equipment, communication services, data processing and system integration. According to Machina Research and Nokia, the revenues of the global

Industrial IoT market will reach 484 billion euros in 2025, and the major players will be the transportation, housing, health and applications for the "Smart homes". Meanwhile, the biggest earnings will be generated by the applications, analytics and services for end users. At the same time, Machina Research and Cisco's overall assessment of the global IoT market (user and corporate) is estimated to reach $4.3 trillion in 2025. The market of smart home technologies is estimated to account for 25 billion euros, which, according to the current exchange rate, equals to more than 1.7 trillion rubles [7].

2.3 Big Data

Big data is a new generation technology designed for cost-effective extraction of useful information from complex or very large amounts of data, using very high speed of collection, processing and analysis [8].

There are many definitions of a threshold beyond which data is considered really large, from petabytes (10^{15} bytes) to exabytes (10^{18} bytes); however, methodologically, the more correct way to describe this scale is to apply the so-called V-model. As its factors are most often use Variety, Velocity (processing speed) and Volume (storage volume). Some sources supplement the model with such factors as Value (data value) and Validity (data reliability), Veracity (data accuracy), and many others [9].

According to IDC research, the global big data market is growing steadily: in 2015 it reached $122 billion, in 2016-already $130 billion, by 2020 the analysts predict the growth to reach $203 billion. Based on internal revenue growth indicators, experts also positively assess the dynamics of the Russian market. Meanwhile, the impact of the global crisis is gradually decreasing [8].

The Russian market forecast for the upcoming years is as follows: the market volume in 2015 will reach $500 million and in 2018 – $1.7 billion. The Russian share in the world market will be about 3% in 2018; the amount of accumulated data in 2020 will be 980 exabytes; the volume of data will grow up to 2.2% in 2020. The most popular technologies in Big data processing will be data visualization, analysis of media files and the IoT [10].

From the results of the analysis, we can conclude that in the near future we will observe the growth of the Russian big data market, accompanied by the expansion of big data processing technologies [11].

2.4 Edge, Fog and Cloud Computing

Cloud computing is a distributed data processing technology where the computer resources and capacities are provided to a user as Internet services. Edge computing is a new computing model that allows to store and (or) process data at the edge of a network and provide intelligent services near the data source in conjunction with "cloud" computing. Fog computing involves the transfer of computing and data storage from a traditional cloud to an intermediate layer of devices located closer to the edge of a network, which reduces the load on the communication medium and devices [8, 9].

There exist several types of architectures that can be built on cloud computing. They involve the distributed logic of data processing - so called preprocessing. With the increase in the amount of data transmitted and the use of IoT, the following problems arise [8, 9]:

- The IoT data is stored in huge volumes and this data must be continuously analyzed; the interconnections, inconsistencies and discrepancies must be identified, i.e. the calculations with the linear growth of centralized cloud computing do not meet the needs of data processing from multiple data sources;
- The network bandwidth and data rates have become a weak spot as the number of users increased (data generated by IoT devices is projected to exceed 847 zettabytes by 2021). We need to keep in mind that the IP-traffic of the global data-processing center can only support 19.5 zettabytes;
- The majority of end users at the edge of the network are typically mobile devices that do not have sufficient computing resources to store and process large amounts of information, and are limited by the battery life;
- As a result of long-distance data transfer and outsourcing, data security and privacy have become a rather complex tasks in cloud computing. Processing data at the cloud edge can reduce the risk of information leakage.

In order to solve these problems, we need to use a combination of either an edge and cloud computing, or fog and cloud computing, or edge, fog and cloud computing.

The main idea of using fog computing is to filter and process data before it reaches the cloud server, i.e. to use local computing power without stopping interaction with cloud services.

The basic idea of using edge computing is to filter and process data using the devices that are as close to engineering objects as possible. Currently, such devices have "weak" intellectualization.

The concurrent use of edge, fog and cloud computing is appropriate when the size of a system is very large (tens of millions of devices) and the network infrastructure extends over thousands of square kilometers.

According to expert statistics, in 2017, Russian domestic market of cloud services increased by 49% and amounted to 663.74 million of US dollars. At the same time, the largest market share of cloud services—61%—belonged to SaaS (Software as a Service); IAAS (Infrastructure as a Service) and PaaS (Platform as a Service) accounted to 30% and 8% respectively. IDC expects PaaS to become the fastest growing category in the next few years [13].

"Russia Cloud Services Market: 2018–2022 Forecast" and The 2017 Analysis, published by IDC, notes that public clouds are still in the highest demand. Microsoft, a leader in this category, has 11% of the market. The largest group of consumers of cloud services are still service-oriented business sectors: in the first place—retail and wholesale; financial and manufacturing sectors took second and third places, respectively [13].

Fog computing is a new stage in the development of cloud computing. It reduces delays that occur when data is transferred to the central cloud and creates new methods of developing intelligent IoT devices. The advantage of fog computing is the decrease in the amount of data transferred to the cloud, which reduces network bandwidth requirements, increases data processing speed and reduces decision delays [13]. Fog computing solves many of the most common problems, such as high network latency, difficulties associated with endpoint mobility, loss of communication, high bandwidth costs, unexpected network congestion, vast geographical distribution of systems and customers.

By 2022, the global market of Fog systems is estimated to reach $18 billion. The greatest potential for the development of fog computing technologies are in the following sectors: energy, utilities, transport, agriculture, trade, health and manufacturing industry.

The energy and utility sectors represent the largest market for Fog computing systems. By 2022, their growth potential is estimated to reach up to $3.84 billion.

Edge calculations are a very promising segment of the market. Gartner anticipates that by 2020, the IoT will include more than 20 billion devices. Approximately 6 billion of them will be used to improve efficiency in business, health, science and public sectors. Rapid growth in the quantity of devices is accompanied by an even more rapid increase in the volume of data created. Gartner expects that by 2020, when the number of IOT devices reaches 20 billion, the amount of generated data will reach 35–45 zettabytes (ZB). By 2025, this number is expected to grow up to more than 150 ZB [7].

By 2023, the world market of edge computing is estimated to grow up to $4.6 billion. Market growth assures the extensive integration of the IoT and virtual and augmented reality into everyday life. CNews also predicts the transition to a new communication standard - 5G. "The World Market for Edge Computing – Predictions for 2018–2023" published on Researchandmarkets.com agrees with the statements of CNews [27]. Gartner believes that edge computing is one of the top 10 technology trends of 2019.

2.5 Data Mining

Data mining is the process of identifying hidden patterns or relationships between variables in large quantity of raw data. It is divided into classification, modeling, forecasting and other tasks. Data mining is based on methods and models of statistical analysis and machine learning with special emphasis on automatic data analysis [5].

When developing monitoring systems of housing and utility services, we strive to create devices that use methods and tools of weak artificial intelligence in order to analyze object parameters and set up a two-way communication with the system of monitoring and control.

In 2017, Gartner identified ten strategic trends that would stimulate the development of four major path of strategic development in organizations. The first path, "Intelligence Everywhere" covers technologies and methods of data processing, which include advanced machine learning and artificial intelligence. They enable us to create intelligent hardware and software systems that can self-learn and adapt. The second path involves technologies focused on the ever closer links between the real and digital world. The third path is an integration of platforms and services necessary for merging of intelligent digital technologies. The fourth one encompasses all aspects of adaptive security architecture [14].

According to Gartner's forecast for 2019–2020, 80% of artificial intelligence products will continue to be akin to alchemy and only a small contingent of dedicated people will work with them. Lack of necessary experience and skills will remain a major barrier to the widespread adoption of AI products. As for the near future, by the end of 2019, research on automation of intellectual data processing will transcend the growth of AI complexity and the shortage of professionalism will begin to decline.

As the market for business intelligence (BI) products is growing, the increase in development of methods used for intellectual processing of semi-structured data is becoming very prospective. For example, the revenue growth of the largest Russian BI companies I-Teco, RDTEX, Prognoz, CROC, HeliosITin 2013–2014 went up to 28%; in some cases (FORS, BARS Group) it reached 60%. According to IDC data for April 2013, global spending on BI services increased by an average of 14.3%; and, in 2016, it amounted to $ 70.8 billion.

Just as Gartner has foreseen, the BI systems and analytical platforms market remained one of the fastest growing segments of the global software market until 2016. The average annual growth rate was about 7% from 2011 to 2016. By 2016; the market volume has reached $17.1 billion. The BI market coupled with data warehouses was growing at the rate of 9% per year [14].

3 The Proposed Approach

The development of an information system for monitoring the engineering objects of housing and utility services is fully based on the digital technology we have discussed above. The functions of the system include the following [9]:

- Fault Detection –monitoring of equipment condition and fault detection;
- Fault Classification-determining the cause of a malfunction;
- Fault Detection and Classification – concurrent monitoring of equipment condition and determining the cause of the malfunction;
- Fault Prediction - predicting the occurrence of malfunctions;
- Equipment Health Monitoring - detection of anomalies and deviations from standard equipment operations;
- Predictive Maintenance - setting the best time for equipment maintenance based on the data of its operations.

The system architecture allows the implementation of distributedlogic. When working with data, typical schemes of the IIoT employ both, hot (online) data processing and cold processing.

The developing information system of monitoring uses hot processing on the basis of edge computing. This involves minimal conversions and transformations necessary for the next steps: sensor readings, local analysis in real time and filtering. Hot processing is performed on the basis of devised association rules [15] and implemented in form of regulating and monitoring controller, which is patented [16].

Cold processing is implemented on the basis of cloud computing and includes data storage, knowledge extraction based on data mining, reporting, predictive analytics and decision making. Predictive analysis in the monitoring system is carried out using Markov chains [17] and cluster analysis [18]. Figure 2 shows our modification of the FP-growth algorithm used to form Association rules.

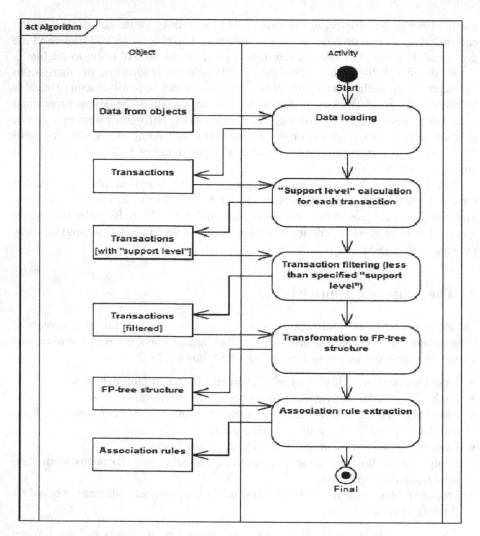

Fig. 2. Application of the FP-growth algorithm to form associative rules

The conceptual diagram of interaction among the developing information system, private household objects and users is shown in Fig. 3. The diagram consists of three interconnected blocks – the objects of housing and utilities complex; information system and users.

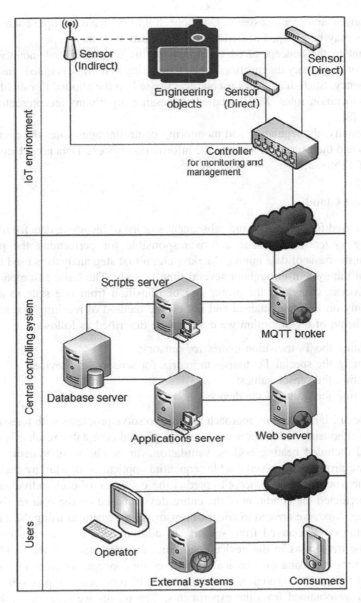

Fig. 3. Conceptual diagram of information system and external user interaction

3.1 Objects of the Housing and Utilities Complex

This unit collects data from the engineering systems, preprocesses and transfers it to the information system. The condition of the engineering system is determined using special sensors that regularly monitor certain indicators that affect the operation of the engineering system. Each sensor is attached to an individual controller that collects data from the sensor and preprocesses it. One household may have multiple controllers and the sensors may be grouped according to their location or their type.

Pursuant to the concept of edge computing, the regulating and monitoring controller performs primary diagnostics of the engineering system: scheduled, unscheduled and emergency. Such diagnostics is conducted based on the allotted thresholds and the devised association rules. More detailed processing algorithms are presented by the authors in [8].

Additionally, the regulating and monitoring controller aggregates the data from all controllers and their further transfer to the information system. Data is exchanged using the MQTT Protocol.

3.2   Object Cloud

This unit is used for processing and subsequent storage of incoming data from all of the engineering systems. The same unit is responsible for performing the predictive analysis on the basis of data mining. Markov chain first step analysis is used to predict the state of the system throughout several time periods. The basic assumption of the Markov process states that the probability of transition from one state to any other depends only on the state attained and not on the method of reaching that state.

The scheme of the algorithm we used can be described as follows:

- Calculating the Ts transition matrix for sensors;
- Calculating the special TU transition matrix for sensors and devices;
- Forecasting the sensor values;
- Forecasting the status of the device.

The use of Markov chains approach in order to solve problems with forecasting the conditions of housing and utility service devices yielded acceptable results. The devices we studied included heating boilers, ventilation, air conditioning systems, etc. The approach we proposed involved double sequential application of Markov chains. First, we used the Markov chain concept to predict the condition of each individual sensor; then, we predicted the condition of the entire device based on the data received from the first step. Next, we created matrices of sensor state transitions using the analysis of actual readings we received from sensors. In addition, we developed a matrix that reflected the transitions in the device conditions that depended on the conditions of individual sensors. Using the classical Markov chain concept, we were able to predict the state of the system throughout several periods of time and compare our results to actual data we obtained from the experiments. The results we received in this paper were used in the implementation of forecasting functions for predictive maintenance of actual real equipment in "smart homes".

Figure 4 shows our predictive Analytics algorithm based on Markov chains. More detailed results are described in [15, 17, 18].

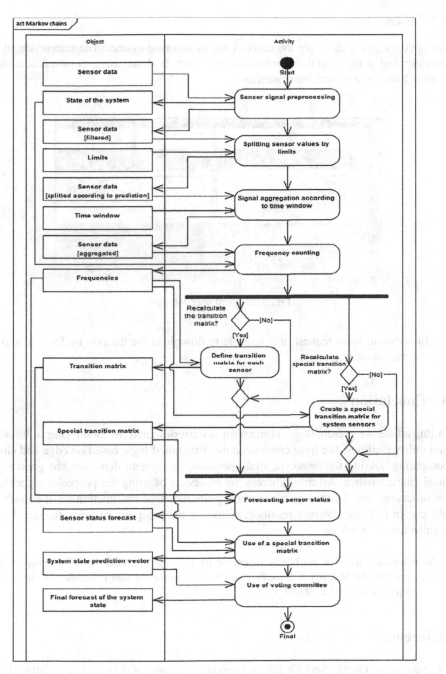

Fig. 4. Markov chain-based algorithm for predictive analytics

3.3   Users

An operator and a client are the users of the monitoring system. The interaction of an operator and a client and the information system is done using personal accounts. Figure 5 shows a typical user interface.

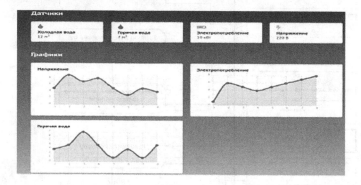

Fig. 5.   Typical user interface

In addition, upon request, it is feasible to download the data (in uniform mode) to the external storage systems.

4   Conclusion

In this article we presented an information system designed for monitoring of housing and utility facilities. We have considered the distributed logic based on edge and cloud computing. During the prototype trial operation, the system demonstrated good technical characteristics. All this indicates the prospects of using the proposed algorithms for processing and forecasting of emergency situations, of malfunctioning of household objects. In fact, the obtained results demonstrate the prospects of using the developed architecture as a whole.

Acknowledgement. This work was supported by the MEPhI Academic Excellence Project (agreement with the Ministry of Education and Science of the Russian Federation of August 27, 2013, project no. 02.a03.21.0005).

References

1. Newsroom: Gartner Says 6.4 Billion Connected "Things" Will Be in Use in 2016, Up 30 Percent From 2015. Gartner, Inc. (2017). http://www.gartner.com/newsroom/id/3165317
2. Tadvisor: Government Business IT. https://www.tadviser.ru/index.php/IoT,_M2M
3. Sethi, P., Sarangi, S.R.: Internet of things: architectures, protocols, and applications. J. Electr. Comput. Eng. **2017**, 9324035 (2017). https://doi.org/10.1155/2017/9324035

4. Stojkoska, B.R., Trivodaliev, K., Davcev, D.: Internet of things framework for home care systems. Wirel. Commun. Mob. Comput. **2017**, 10 (2017). https://doi.org/10.1155/2017/8323646. 8323646

5. Tupchienko, V.A.: Digital Platforms for Managing the Life Cycle of Complex Systems. Nauchnyj konsul'tant, Moscow (2018)

6. Robosapiens. https://robo-sapiens.ru/stati/umnyiy-gorod/

7. Vlasova, Y.E., Kireev, V.S.: Overview of the Russian market of IOT-technologies. Mod. Sci.-Intensive Technol. **8**, 48–53 (2018)

8. Kireev, V.S., et al.: Predictive repair and support of engineering systems based on distributed data processing model within an IoT concept. In: Proceedings -2018 6th International Conference on Future Internet of Things and Cloud Workshops, pp. 84–89 (2018)

9. Kireev, V.S., Filippov, S.A., Guseva, A.I., Bochkaryov, P.V., Kuznetsov, I.A., Migalin, V.: Cloud computing in housing and utility services monitoring systems. In: Proceedings -2018 6th International Conference on Future Internet of Things and Cloud Workshops, pp. 90–94 (2018)

10. Moyne, J., Iskandar, J.: Big data analytics for smart manufacturing: case studies in semiconductor manufacturing. Processes **5**, 39 (2017)

11. Barman, A., Ahmed, H.: Big Data in Human Resource Management Developing Research Context (2016). https://doi.org/10.13140/rg.2.1.3113.6166

12. Phan, A.M., Nurminen, J.K., Francesco, M.Di.: Cloud databases for internet-of-things data. In: 2014 IEEE International Conference on Internet of Things (iThings), and Green Computing and Communications (GreenCom), IEEE and Cyber, Physical and Social Computing (CPSCom), pp. 117–124 (2014). https://doi.org/10.1109/ithings.2014.26

13. Tadvisor: Government Business IT. http://www.tadviser.ru/index.php/28Fog_computing%29

14. RBC Group. http://www.rbcgrp.com/files/QlikView_TAdviser2013.pdf

15. Kireev, V.S., Guseva, A.I., Bochkaryov, P.V., Kuznetsov, I.A., Filippov, S.A.: Association rules mining for predictive analytics in IoT cloud system. Adv. Intell. Syst. Comput. **848**, 107–112 (2018)

16. Vlasov, A.I., Echeistov, V.V., Krivoshein, A.I., Shakhnov, V.A., Filin, S.S., Migalin, V.S.: An information system of predictive maintenance analytical support of industrial equipment. J. Appl. Eng. Sci. **16**(4), 560, 515–522 (2018)

17. Kireev, V.S., Guseva, A.I.: Application of Markov chains for the solution of the predictive maintenance tasks in the fields of IoT and housing and utility services management. In: Proceedings of the XX International Conference "Data Analytics and Management in Data Intensive Domains" (DAMDID/RCDL 2018), Moscow, Russia, 9–12 October (2018)

18. Bochkaryov, P.V., Guseva, A.I.: The use of clustering algorithms ensemble with variable distance metrics in solving problems of web mining. In: Proceedings -2017 5th International Conference on Future Internet of Things and Cloud Workshops, pp. 41–46 (2017)

K-Means Method as a Tool of Big Data Analysis in Risk-Oriented Audit

Pavel Y. Leonov[1]([⊠]) [iD], Viktor P. Suyts[2]([⊠]) [iD],
Oksana S. Kotelyanets[1]([⊠]) [iD], and Nikolai V. Ivanov[1]([⊠]) [iD]

[1] Department of Financial Monitoring, National Research Nuclear University
MEPhI (Moscow Engineering Physics Institute), Moscow, Russia
{pyleonov, OSKotelyanets}@mephi.ru,
ivanov.nikolay.7ll@gmail.com
[2] Department of Accounting, Analysis and Audit of the Economic Faculty,
Lomonosov Moscow State University, Moscow, Russia
viktor.suyts@gmail.com

Abstract. Considering the modern risk-oriented approach to auditing, environmental instability, as well as the lack of clear recommendations for conducting selective surveys, improving sampling methods is relevant, updating and new development of methodological tools for conducting selective audits are required; the article substantiates the use of the K-means method as a selective method for constructing an audit sample, special attention was paid to the professional judgment of the auditor and his need to apply K-means clustering.

Keywords: Big data analysis · K-means method in audit · Forensic · Compliance

1 Introduction

With the sharp (due to the emergence of associations of organizations) growth of the volume of information contained in the accounting (financial) statements, the use by auditors of a complete method of verifying reports has become impossible. As a result, there is a need to apply sampling methods, because testing on a sample basis is the most expeditious method of conducting an audit and allows making timely decisions on eliminating the deficiencies of the internal control system, and also provides an opportunity to reduce the level of audit customers.

2 K-Means Method as a Tool for Building an Audit Sample

Cluster analysis is a multidimensional statistical procedure that performs the collection of data containing information about a set of objects, followed by the ordering of objects into relatively homogeneous groups. The work of cluster analysis is based on two principles [1]:

- considered features of an object allow splitting of a set of objects into clusters;
- correctness of the choice of scale or units of measurement of signs.

© Springer Nature Switzerland AG 2019
M. Younas et al. (Eds.): Innovate-Data 2019, CCIS 1054, pp. 206–216, 2019.
https://doi.org/10.1007/978-3-030-27355-2_16

The task of cluster analysis is to split a set of objects $O = \{O_1, O_2, \ldots, O_n\}$, statistically represented as a matrix "object-property" (1), into relatively small (previously known or not) homogeneous groups or clusters.

$$X_{n*p} = \begin{pmatrix} x_{11} & x_{12} & \cdots & x_{1p} \\ x_{21} & x_{22} & & x_{2p} \\ \vdots & & \ddots & \vdots \\ x_{n1} & x_{n2} & \cdots & x_{np} \end{pmatrix} \tag{1}$$

where n – number of objects;

   p – number of properties describing the object.

Depending on the volume n of classified observations and on a priori information on the number of clusters, the clustering methods are divided into hierarchical and non-hierarchical (iterative) (see Fig. 1).

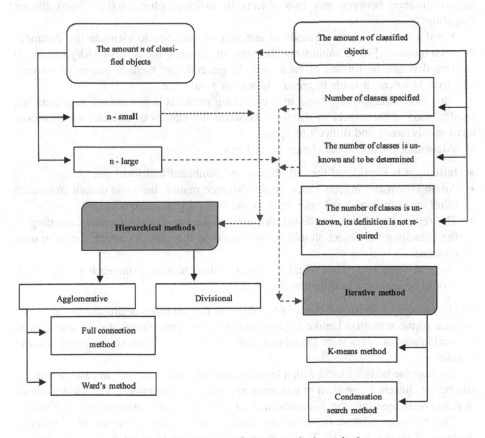

Fig. 1. Classification of cluster analysis methods

Hierarchical cluster procedures do not depend on a priori information on the number of clusters and are used when the number of observations n is small. The hierarchical clustering methods are divided into agglomerative and divisional [2].

The principle of operation of hierarchical agglomerative (divisional) procedures consists in the sequential unification (separation) of groups of objects, first the closest (far) and then more and more distant from each other (close to each other) [3].

Algorithm for agglomerative cluster procedures:

- Centering and rationing the values of the original signs;
- Calculation of the matrix of distances between objects. Every object is a class;
- Combining the two closest classes;
- Recalculation of the matrix of distances between classes of objects;
- Repeat steps 3–4 until all objects are combined into one class.

Among the methods of agglomerative cluster analysis are the following:

The full connection method - the distances between clusters are determined by the largest distance between any two objects in different clusters (i.e. "most distant neighbors");

Ward's method - uses methods of analysis of variance to estimate the distances between clusters [4]. It minimizes the sum of squares for any two (hypothetical) clusters that can be formed at each step. In general, the method seems to be very effective, however, it tends to create clusters of small size.

The disadvantage of agglomerative clustering methods is that at each step there is a recalculation of the matrix of distances between the classes of objects, which makes them cumbersome and difficult to apply.

Algorithm for divisional cluster procedures:

- Initially it is considered that all objects are combined and form one cluster;
- Among the many objects based on the distance matrix, the most distant from each other are determined and take them as the basis for two new clusters;
- The remaining objects are distributed among the two classes formed according to the principle: the object should be attributed to the class to which the minimum distance.
- Then in these two classes find the most distant objects from each other, which should be assigned to different classes, etc.

The advantage of divisional cluster procedures is that all calculations are based on the original distance matrix. Unlike agglomerative cluster procedures, it does not need to be recalculated at every step. Divisional methods of cluster analysis are rarely used in practice.

Iterative methods are used with a large amount of n observations, in cases where the number of clusters is specified or unknown and must be determined. The main methods of iterative analysis are the K-means method and the condensation search method.

The K-means method is one of the most common and frequently used iterative methods of cluster analysis. This method was invented in the 1950s by the mathematician Hugo Steinhaus and almost simultaneously by Stuart Lloyd. After the work of McQueen, he gained particular popularity, therefore this method is also called the McQueen method.

Unlike hierarchical methods, which do not require preliminary assumptions about the number of clusters, to be able to use this method, it is necessary to have a hypothesis about the most probable number of clusters. The K-means method builds exactly k different clusters located at large distances from each other. Calculations begin with k randomly selected observations that become the centers of groups, after which the object composition of the clusters changes to minimize variability within the clusters and maximize the variability between clusters.

The mechanism of the K-means method is fairly simple. A certain set consisting of n objects is given, which should be divided into k clusters [5]. The division of a set of objects into clusters occurs in several stages:

From the existing totality are selected randomly or specified by the researcher based on any a priori considerations of k objects, where k is a given number of classes. These k objects are taken as class ε standards. Each standard is assigned a sequence number, which is the cluster number.

Of the remaining (n − k) objects, each object is extracted and checked, to which of the standards it is closest.

The most common methods for calculating the distance between objects and standards are:

- Euclidean distance.

$$d_E(O_i, O_j) = \sqrt{\sum_{l=1}^{m} (x_{il} - x_{jl})^2} \tag{2}$$

- Minskovsky distance.

$$d_M(O_i, O_j) = \sqrt{\left(\sum_{l=1}^{m} |x_{il} - x_{jl}|\right)^{1/p}} \tag{3}$$

- Hamming distance (Manhattan distance).

$$d_H(O_i, O_j) = \sqrt{\sum_{l=1}^{m} |x_{il} - x_{jl}|} \tag{4}$$

where O_i и O_j – objects the distance between which we calculate;
      m – number of features describing the object;
      p – any arbitrary number.

After considering each of the (n − k) objects will be divided into k classes - this is the end of the first iteration of the algorithm.

At the next iteration, it is necessary to calculate the distance from each object to the class standard and again distribute n objects into k clusters. In this case, the standard is replaced by a new one, recalculated with regard to the attached object, and its weight (the number of objects included in this cluster) is increased by one. Recalculation of standards and weights is as follows:

$$
\varepsilon_i^\vartheta = \begin{cases} \frac{w_i^{\vartheta-1} * \varepsilon_i^{\vartheta-1} + O_{k+\vartheta}}{w_i^{\vartheta-1}+1}, & if\ d\left(O_{k+\vartheta}, \varepsilon_i^{\vartheta-1}\right) = \min_{1 \le j \le k} d(O_{k+\vartheta}, \varepsilon_j^{\vartheta-1}) \\ \varepsilon_i^{\vartheta-1}, otherwise \end{cases} \tag{5}
$$

$$
w_i^\vartheta = \begin{cases} w_i^{\vartheta-1}+1, & if\ d\left(O_{k+\vartheta}, \varepsilon_i^{\vartheta-1}\right) = \min_{1 \le j \le k} d(O_{k+\vartheta}, \varepsilon_j^{\vartheta-1}) \\ w_i^{\vartheta-1}, otherwise \end{cases} \tag{6}
$$

where $O_{k+\vartheta}$ – selected object to be added to the closest cluster;
ε_i^ϑ – the center of the i-th cluster, after adding a new object;
w_i^ϑ – weight of the i-th cluster, after adding a new object;
$\varepsilon_i^{\vartheta-1}$ – the center of the i-th cluster, before adding a new object;
$w_i^{\vartheta-1}$ – weight of the i-th cluster, before adding a new object.

1. In the next step, one more object is selected. And for him the whole procedure is repeated. Thus, in (n − k) steps, all objects in the aggregate will be assigned to one of the k clusters.
2. To achieve sustainable partitioning, all objects are again attached to the standards obtained after the iteration, while the weights continue to accumulate. Steps 3–4 are repeated.
3. The new partition is compared with the previous one. If they coincide, then these standards are the best solution, otherwise we continue iterations (we return to step 5).

The algorithm is completed when at some iteration there is no change in the intracluster distance. This happens in a finite number of iterations, since the number of possible partitions of a finite set is finite, and at each step the total standard deviation decreases, therefore looping is impossible.

When using different clustering methods for the same set, different splitting options can be obtained (different number of clusters; clusters differ both in composition and in proximity of objects). It is important to remember that the use of the K-means method is possible only when the properties describing objects are numerical in nature.

K-means clustering is used in various areas of human activity: medicine, chemistry, psychology and much more. The author of this method will be used in the audit, namely in the construction of the audit sample [6].

Considering the classification of sampling methods for building an auditory sample, the author proposed to designate the use of the K-means method as a representative non-statistical sample with elements of the non-statistical method of "serial selection". The difference between the K-means method and the "serial selection" method is that when using the "serial selection" method, the total sampling error extends to the population being checked in proportion to the ratio of the volumes of the population being tested and the sampling error, and using the K-means method implies that the total error propagates the sample for the checked population does not depend on the volume of the checked population.

The use of the K-means method in constructing an auditory sample cannot be attributed to statistical methods, despite the fact that this method is mathematical and contains a large number of calculations. The substantiation of this statement for the author was the fact that it is on the basis of his professional judgment that the auditor will choose the following:

- characteristics describing elements of the general population;
- number of clusters into which the population will be divided;
- group of objects for further research using a continuous check, after separation into clusters.

When spreading the total sampling error to the checked population, the author proposed to multiply the total sampling error by two, it is assumed that with the correct choice of object properties, the number of clusters, and after splitting into clusters one/several clusters as an audit sample, at least 50% of errors inherent in the entire population tested.

3 Example of Application of the K-Means Method and Analysis of the Results in Comparison with the Continuous Test

For a more illustrative example of the Application of the K-means method in the audit, the author conducted an audit to identify theft in the sale of petroleum products at gas stations. The audit was conducted in three stages [6]:

1. The preparatory stage (determination of the tested population, the choice of the properties of the object and the number of clusters, the rationale for choosing a cluster (clusters) after applying the K-means method);
2. Application of the K-means method (the stage of calculations using Excel and STATISTICA 10);
3. Analysis of the results obtained (comparison of the results obtained with the results, when conducting a continuous test).

3.1 Preparatory Stage

At the first stage card for account 90 (subaccount 90.1) was downloaded from 1C: Enterprise for a period between 01.01.2017 and 31.06.2017. The sum of operations with journal entries -Dr50, Cr90.1 and Dr57, Cr90.1 were summarized for each day, i.e. the amounts of contracts for the supply and delivery of petroleum products to other organizations were excluded from the daily revenue (transactions with Dr62 Cr90.1). The obtained data was taken as the first characteristic describing n observations $X = (x\_1, x\_2, ..., x\_n)$.

The cluster analysis was carried out relative to the dates of the first half of the year (181 objects), in which the presence of theft of fuel and lubricants by employees was assumed.

According to the authors' research, it was revealed that about 504 sales of petroleum products occur in one day during a 24-hour petrol station operation, thus 90,720 sales of petroleum products occur in 181 days.

As well as properties of dates it is offered to choose:

- X – daily sales of gas stations;
- Y – the share of daily revenue of gas stations in the maximum for the week (formula (7)) (Table 1).

Table 1. A fragment of the initial data used in clustering by the K-means method

Object (date)	X	Y
01.01.2017	212 717,87	0,48
02.01.2017	365 262,68	0,80
03.01.2017	417 150,96	0,81
04.01.2017	441 842,85	0,86
05.01.2017	453 833,89	0,85
06.01.2017	514 804,33	0,96
07.01.2017	446 359,93	0,83
08.01.2017	535 905,47	1,00
09.01.2017	480 512,58	0,90
10.01.2017	500 808,82	0,92
...	...	...

$$y_i = \frac{x_i}{\max_{i-3 \le p \le i+3} x_p}, \text{ where } i = 1, \ldots, 181 \tag{7}$$

The number of clusters was decided to choose k = 4. The author justifies this choice by the content of clusters, namely:

- The days in which the highest revenue was obtained during the current week and the period as a whole;
- Days during which the proceeds were received, which has a slight deviation from the maximum in the current week;
- Days during which the average revenue was obtained relative to the current week and the period as a whole;
- Days with the lowest revenue during the current week and the period as a whole.

The author claims that for further research it is necessary to take the elements of the second cluster. This choice is explained as follows: the likelihood that with a large number of sales per day and having time to perform manipulations with a daily report for a shift, the employee will decide to carry out the theft of fuel and lubricants more than in other cases [7]:

- with a very large number of sales of petroleum products, it is easy to hide their theft, however, workers will hardly have time to spend various frauds on theft of petroleum products, therefore the inclusion of the first cluster in the audit sample is a controversial issue;
- average deviations in revenue from the sale of petroleum products may be associated with both a small number of buyers throughout the day and the influence of the weather (at low temperatures, the amount of fuel and lubricants drops, despite the fact that the mass does not change), therefore dates in the third cluster will not be included in the audit sample;
- large deviations from the maximum revenue are rarely associated with embezzlement, a reasonable employee understands - embezzlement for a large amount has a high level of risk for the employee, since such embezzlement is easy to detect, therefore, research of the fourth cluster objects is unwise.

3.2   Application of the K-Means Method

Given the fact that the selected properties have a different range of values, the first step is to standardize the data. For this, the mean value and the standard deviation of each property were calculated (see Table 2).

Table 2. Mean values and standard deviations of properties

	Mean value		Standard deviation
x_{cp}	612 660,89	σ_X	113 198,53
y_{cp}	0,8873	σ_Y	0,0932

As a result of the standardization of data X and Y (8), new values are obtained: $\widehat{X}$ and $\widehat{Y}$ (see Table 3).

Table 3. A fragment of standardized data used in clustering by the K-means method

Object (date)	$\widehat{X}$	$\widehat{Y}$
01.01.2017	−3,5331	−4,3572
02.01.2017	−2,1855	−0,8854
03.01.2017	−1,7271	−0,8267
04.01.2017	−1,5090	−0,3118
05.01.2017	−1,4031	−0,4343
06.01.2017	−0,8645	0,7870
07.01.2017	−1,4691	−0,5841
08.01.2017	−0,6781	1,2097
09.01.2017	−1,1674	0,1001
10.01.2017	−0,9881	0,3038
...	...	...

$$\tilde{x}_i = \frac{\left(x_i - x_{cp}\right)}{\sigma_X} \text{ and } \tilde{y}_t = \frac{\left(y_i - y_{cp}\right)}{\sigma_Y}, \text{ where } i = 1, \ldots, 18 \qquad (8)$$

The data obtained using Excel was loaded into STATISTICA 10, then the function "Cluster Analysis: k-means clustering" was selected and the number of clusters was set k = 4.

As a result of clustering, 181 objects were divided as follows: 1 cluster - 36 days; 2 cluster – 70 days; Cluster 3 - 42 days; Cluster 4 - 33 days (see Table 4).

Table 4. Fragment of the list of dates for belonging to clusters

Cluster 1	Cluster 2	Cluster 3	Cluster 4
01.01.2017	06.01.2017	12.04.2017	01.02.2017
02.01.2017	08.01.2017	13.04.2017	14.04.2017
03.01.2017	09.01.2017	18.04.2017	15.04.2017
04.01.2017	10.01.2017	23.04.2017	20.04.2017
05.01.2017	11.01.2017	24.04.2017	21.04.2017
07.01.2017	12.01.2017	26.04.2017	22.04.2017
15.01.2017	13.01.2017	27.04.2017	25.04.2017
17.01.2017	14.01.2017	30.04.2017	28.04.2017
21.01.2017	16.01.2017	01.05.2017	29.04.2017
22.01.2017	18.01.2017	02.05.2017	05.05.2017
...	...	...	...

The audit sample included 70 dates belonging to the second cluster. The audit on the identification of theft of petroleum products was as follows:

- Comparison of data on the sale of petroleum products through fuel distribution columns for the types of payments reflected in the final reports for the shift with similar final reports for the shift, printed from the TOPAZ-AZS program;
- Comparison of data on the consumption of petroleum products according to the indications of the program counter of petroleum products passed through the sleeves of the fuel distribution columns, reflected in the final reports for the shift with similar final reports for the shift, printed from the TOPAZ-AZS program.

Instead of comparing different reports for 181 days, a comparison was made of reports for 70 days that were in the audit sample, which already reduced the audit time by 61.33%. According to the results of the comparison, it was revealed that in the 31st day of the 70 belonging to the audit sample, theft of oil products was made.

In order to analyze the effectiveness of the application of cluster analysis by the K-means method, a full-scale test was performed, i.e. Comparison of reports was made for all 181 days. According to the results of the continuous check, it was revealed that in 50 days of the selected period, theft of fuel and lubricants was carried out by employees of petrol stations. The identified days of fraud were considered to belong to the K-medium clusters obtained by the method (the results are presented in Table 5 and Figs. 2 and 3).

Table 5. Belonging of days of theft to the received clusters

Cluster's No.	The total number of days	Number of days of theft
I	36	10
II	70	31
III	42	4
IV	33	5
Total	181	50

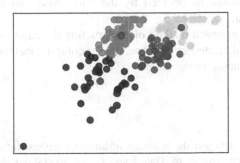

Fig. 2. Cluster visualization

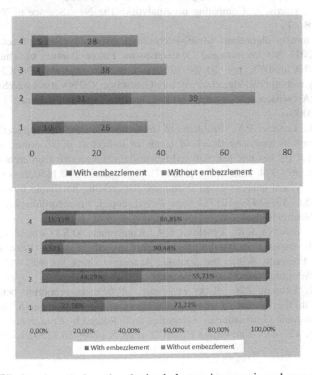

Fig. 3. Affiliation days theft to the obtained clusters in numeric and percentage terms

According with chosen method or tool of risk-based audit depends on the result of the test, and the resources spent on its conduct. Therefore, the advantage of the K-means method is the speed and ease of implementation.

The analysis of the obtained results made it possible to conclude that when using the K-means method in the construction of the audit sample during the audit to identify theft of fuels and lubricants at gas stations, the duration of the audit will be reduced by almost three times (by 61.33%), while 62% of the dates in which the theft was made will be revealed.

The K-means method serves as a tool to facilitate the division of data into groups, but how to use it should be decided by the audit based on their experience and professional judgment.

The result of the presented method of constructing the audit sample is the rationale for the development of clustering methods, in particular the method of K-means, in the construction of the sample in the audit.

References

1. The analysis of big data and the accuracy of financial reports. In: 2017 5th International Conference on Future Internet of Things and Cloud Workshops (FiCloudW), pp. 53–56 (2017). https://doi.org/10.1109/FiCloudW.2017.93, http://gateway.webofknowledge.com/gateway/Gateway.cgi?GWVersion=2&SrcAuth=Alerting&SrcApp=Alerting&DestApp=WOS&DestLinkType=FullRecord;UT=WOS:000435144700010

2. Davenport, T., Harris, J.: Competing on Analytics: The New Science of Winning (2010). BestBusinessBooks

3. Use of data mining algorithms while determining threshold values of economic security indices. In: 2017 5th International Conference on Future Internet of Things and Cloud Workshops (FiCloudW), pp. 20–24 (2017). https://doi.org/10.1109/FiCloudW.2017.73, http://gateway.webofknowledge.com/gateway/Gateway.cgi?GWVersion=2&SrcAuth=Alerting&SrcApp=Alerting&DestApp=WOS&DestLinkType=FullRecord;UT=WOS:000435144700004

4. Zhukov, A.N., Leonov, P.Y.: Problems of data collection for the application of the data mining methods in analyzing threshold levels of indicators of economic security. In: III Network AML/CFT Institute International Scientific and Research Conference "FinTech and RegTech", pp. 369–374. Knowledge E (2017). https://doi.org/10.18502/kss.v3i2.1566

5. Leonov, P.Y.: Visual analysis as an instrument for creating unreliable financial statements. Sci. Vis. **9**(5), 137–143 (2017). https://www.scopus.com/record/display.uri?eid=2-s2.0-85039864252&origin=resultslist

6. Leonov, P.Y., Leonova, E.K., Veselova, E.A., Kotelyanets, O.S.: General scheme of risk – oriented audit stages. In: Breakthrough Directions of Scientific Research at MEPhI: Development Prospects Within the Strategic Academic Units, pp. 402–415. Knowledge E (2018). https://doi.org/10.18502/keg.v3i6.3021

7. Shadrin, A.S., Leonov, P.Y.: Risks evaluation of financial-economic activity and their management in the system of economic security of the organization. In: III Network AML/CFT Institute International Scientific and Research Conference "FinTech and RegTech", pp. 427–435. Knowledge E (2017). https://doi.org/10.18502/kss.v3i2.1573

Author Index

Ait Wakrime, Abderrahim   47

Blahuta, Jiri   150
Bochkaryov, Pyotr V.   193
Boubaker, Souha   47

Crump, Trafford   92

Demir, Onur   142
Di Modica, Giuseppe   18
Dogan, Kadriye   32
Dronyuk, Ivanna   121

Far, Behrouz H.   92
Filippov, Stanislav A.   193
Flunger, Robert   133
Fountouris, Antonis   3

Gaaloul, Walid   47
Greguš ml., Michal   121
Guseva, Anna I.   193

Incel, Ozlem Durmaz   32
Ivanov, Nikolai V.   179, 206
Izonin, Ivan   121

Kallel, Slim   47
Kireev, Vasiliy S.   193
Kocak, Berkay   142
Kondylakis, Haridimos   3

Kotelyanets, Oksana S.   179, 206
Kuznetsov, Igor A.   193

Leonov, Pavel Y.   179, 206
Liu, Ningning   65

Miloslavskaya, Natalia   165
Mladenow, Andreas   133
Mohammed, Emad   92

Nikiforov, Andrey   165

Onan, Aytuğ   80, 107

Peng, Shaowen   65
Plaksiy, Kirill   165
Planas, Apostolos   3
Plexousakis, Dimitris   3

Sharifi, Fatemeh   92
Soukup, Tomas   150
Strauss, Christine   133
Su, Chang   65
Suyts, Viktor P.   179, 206

Tkachenko, Pavlo   121
Tkachenko, Roman   121
Toçoğlu, Mansur Alp   107
Tomarchio, Orazio   18
Troullinou, Georgia   3

Xie, Xianzhong   65

Printed in the United States
By Bookmasters